Analytics in Healthcare and the Life Sciences

Analytics in Healthcare and the Life Sciences

Strategies, Implementation Methods, and Best Practices

Edited by Dwight McNeill
Foreword by Thomas H. Davenport

Vice President, Publisher: Tim Moore
Associate Publisher and Director of Marketing: Amy Neidlinger
Executive Editor: Jeanne Glasser Levine
Developmental Editor: Natasha Torres
Operations Specialist: Jodi Kemper
Cover Designer: Alan Clements
Managing Editor: Kristy Hart
Project Editor: Elaine Wiley
Copy Editor: Geneil Breeze
Proofreader: Debbie Williams
Indexer: Angela Martin
Compositor: Nonie Ratcliff
Manufacturing Buyer: Dan Uhrig

For information about buying this title in bulk quantities, or for special sales opportunities (which may include electronic versions; custom cover designs; and content particular to your business, training goals, marketing focus, or branding interests), please contact our corporate sales department at corpsales@pearsoned.com or (800) 382-3419.

For government sales inquiries, please contact governmentsales@pearsoned.com.

For questions about sales outside the U.S., please contact international@pearsoned.com.

Printed in the United States of America

First Printing November 2013

ISBN-10: 0-13-340733-0
ISBN-13: 978-0-13-340733-4

Pearson Education LTD.
Pearson Education Australia PTY, Limited.
Pearson Education Singapore, Pte. Ltd.
Pearson Education Asia, Ltd.
Pearson Education Canada, Ltd.
Pearson Educación de Mexico, S.A. de C.V.
Pearson Education—Japan
Pearson Education Malaysia, Pte. Ltd.

Library of Congress Control Number: 2013949115

To all the people pioneering the field of analytics in healthcare.

Contents

Foreword

What's more important to healthcare than analytics? We can use this powerful resource to determine what treatments are most likely to be effective, which care practices are worth the cost, which patients deserve special attention, and which of those patients are likely not to take their medications. In a country—and to a lesser degree, a world—in which we spend too much on healthcare and get too little in return, analytics can help to restore a balance between cost and value. And given all the possible things that we can do with analytics, what could be more important than improving the healthcare of human populations?

For better or worse, however, healthcare is behind other industry sectors in terms of its analytical sophistication. Less charitable observers have told me they think that healthcare is about 25 years behind; I would be more inclined to say five or ten years behind most industries. In any case, healthcare was late to adopt analytics, and late to put in place the data, systems, and skills to use analytics effectively. Other industries have functional silos across which they don't share data and analyses well, but the divide among clinical, operational, and business groups within healthcare organizations leaves other silos behind.

However, when Jack Phillips—the CEO and co-founder with me of the International Institute for Analytics (IIA)—and I began speaking in 2010 with healthcare providers and payers, and life sciences firms, about undertaking sponsored research on healthcare analytics, the responses couldn't have been more encouraging. We discovered a great hunger for information about how to use analytics more effectively in every aspect and segment of the industry.

Perhaps we shouldn't have been surprised by this positive reception, because other actions being taken by healthcare industry members are consistent with a strong interest in analytics. One important signal of interest, for example, is the widespread activity among healthcare providers in installing, replacing, or updating basic transaction systems. You can't do analytics without solid underlying data on patients, care provision processes, and costs, so provider institutions are pouring massive amounts of resources into electronic health record, billing, and operational management systems. Each of these systems generates new data that only reporting and analysis can make sense of. The key, of course, will be moving beyond the transactional focus to understanding what all the data mean and how analytics on them translate into better patient care.

There is also huge interest across the industry in health data access: patient portals, mobile patient access to records, telemedicine of various types, and the e-patient movement in general. Patients have an interest in seeing their basic transaction information, but they have an even greater interest in analyses and recommendations for personalized care. Only analytics and decision rules can provide this kind of information to caregivers and patients.

Personalized genetic medicine is still on the horizon, but as the cost of sequencing a human genome continues to fall, before long there will be an incredible amount of genetic information available to correlate with disease states and treatment outcomes. Analytics are the only method possible for reducing the mass of data to a comprehensible level, and for understanding the relationships between a particular genome and the care interventions that will improve that patient's health.

Finally, if these indicators weren't enough, there is also a high degree of startup formation in the healthcare big data space—companies dealing with less structured forms of health data. In my hometown of Boston, in particular, there are more than fifty big data startups in a variety of health industry sectors and problem domains. Most of these organizations have venture capital funding; some have already been acquired by larger organizations. Such a big data "gold rush" suggests that there will be many future innovations and potential breakthroughs in the healthcare analytics space.

Healthcare analytics activity also takes place across a variety of sub-sectors within the overall industry, including payers, providers, and life sciences firms. Addressing the content originating in each of these sectors is critical to the focus and value of this book. I've already mentioned that healthcare organizations are siloed with respect to internal functions that generate analytics. However, there is also an extreme lack of integration across sectors. Payers, providers, and life sciences firms don't exchange analytics or even raw data very often.

This situation must change if societies are going to deliver effective and cost-efficient healthcare. Both within and across organizations, the healthcare analytics of the future need to cross boundaries.

There are trends in motion—not only the widespread adoption of electronic medical record systems, but also the "meaningful use" requirements for these systems for reimbursement, and the shift to "accountable care organizations"—that will require greater levels of analytical integration within and across organizations. For the most part, however, the trends have yet to yield dramatic change in analytical integration.

Although we don't have anything resembling full analytical integration in healthcare, it's useful to think of what it might look like. Within provider organizations, for example, integrated analytical decisions about patient care would address clinical, financial, and quality concerns—all at the same time. Care providers would be able to employ clinical decision support tools to administer the most effective treatment protocols, but would also simultaneously understand the financial implications of different treatment approaches. On patient admission, hospitals would understand how likely they were to improve the patient's condition, what the treatment would be likely to cost, and how likely the patient was to be able to pay for the treatment. On discharge, hospitals would know the likelihood of readmission, and the best combination of home healthcare and other interventions to prevent readmission. Primary care institutions would share a patient's data and analytics—using proper privacy protection, of course—with other institutions and individuals that provide care for the patient.

In terms of planning for new services and facilities, providers would have accurate statistical forecasts of patient demand for existing and planned offerings. They would market those offerings to patients most likely to require them. They would understand the implications of new and enhanced service offerings—and the quality with which they are delivered—for the institution's financial and operational performance.

In such an integrated world, payer organizations would take the lead in disease management programs driven by analytics. They would use data about their customers and claims in order to understand what genetic, physiological, and behavioral attributes are associated with particular diseases. After informing their customers or members of any diseases they are likely to contract (assuming patients opt into receiving this information), they would also have analytics on which intervention strategies are most likely to yield the desirable behavior change necessary to avert the disease. They would supply these analytics, again with the appropriate levels of privacy, to anyone who cares for the patient.

Payers would also use analytics to identify employers and providers they want to work with, and who would be likely to employ their services. Payers in the U.S. would also have considerable information on consumers (which will become primary health insurance purchasers under U.S. healthcare reform), and would be able to do predictive modeling of which consumers would be most likely to purchase certain types of insurance.

This integrated vision would also encompass life sciences firms—including pharmaceutical and medical device organizations—which would offer predictive models of responses to drugs and devices. With the advent of personalized genetic medicine, life sciences firms would be able to help care providers understand whether particular drugs and treatment protocols would be likely to work on particular individuals. This would also allow an intelligent decision on whether certain medical interventions would be worth the cost. In addition, life sciences firms would have much more effective models of the business value of relationships with physicians (both individually and as members of social and business networks) and provider organizations, and would target marketing and sales resources to the most likely adopters of particular drug and device interventions. Some of the predictive models for physicians would take into account the analytical results from clinical trials and large-scale population studies.

In addition to these integrated analytics initiatives within healthcare organizations, integration would also feature a variety of integration activities for analytics across sub-sectors of healthcare. In this ideal environment, providers, payers, life sciences firms, pharmacy benefit managers, patient registries, and other organizations would share data and analytics with other organizations within their sector and outside of it. Payers, for example, could share analytics about "at risk" status of their customers with the providers who would treat them for it. All parties would share data and analysis on post-market surveillance of drugs and medical devices.

This is an appealing vision of integration. However, both within and across healthcare industry sectors, analytical integration is still in its infancy. Fortunately, economic and regulatory trends in the industry are beginning to lead to efforts to combine and share data and analytics across organizations and sectors. But for substantial progress to take place, healthcare organizations need to finish implementing basic transaction data systems. They need to create groups whose function it is to integrate and coordinate analytics within and outside the organization. And because these efforts will require investment, advocates for analytical integration need to work closely with senior executives to help them understand the need for and potential of analytics across boundaries.

The integration of chapters in this book is itself a reflection of the integration needed in healthcare analytics. The book is edited by Dwight McNeill, an expert on healthcare analytics and IT, and a personal force for integration. He has been a consultant, entrepreneur, professor, regulator, and large company executive in the healthcare information domain. He has been IIA's lead faculty member for our Healthcare Analytics Research Council for several years.

Dwight and I together have authored about half of the chapters in this book. In addition, there are expert authors from consulting firms, analytics and IT vendors, medical centers, and leading practitioners. It would be difficult to imagine a better group of thinkers to address the integration of healthcare analytics within organizations and across payers, providers, and life sciences organizations.

I am confident that readers of the chapters in this book—both those derived from IIA "research briefs" and from "leading practice briefs," or case studies—will find the blueprint and examples for analytical integration within and across organizations. The healthcare organizations that study these materials, recognize the key issues, and create initiatives to address them will be the pioneers in moving us toward a more analytical future in healthcare.

—Thomas Davenport, Cofounder and Research Director of the International Institute for Analytics

About the Authors

Kathleen Aller is the director of data content for McKesson Enterprise Intelligence. Her current interest is the operational and strategic challenges related to automating healthcare quality and operation measures. She has a particular focus on quality eMeasures and on meaningful use measurement for federal electronic health record incentives. She has spoken and written for Healthcare Information and Management Systems Society (HIMSS), Healthcare Financial Management Association (HFMA), and American Medical Informatics Association (AMIA).

John Azzolini has worked in the healthcare field for 30 years as a nurse and healthcare data analyst. In his 18 years at Thomson Reuters, working as both a director of consulting and practice leadership, much of his work has focused on assisting health plans in the assessment of care processes, the creation of metrics that examine the quality of care, employer account reporting, provider profiling, and disease management program evaluation. Mr. Azzolini earned a Master of Business Administration and Master of Public Health from Columbia University and a Bachelor of Nursing from Cornell University.

Albert Bonnema, MD, MPH, Lt. Col., USAF, MC; Air Force Medical Service Deputy CIO.

Deborah Bulger is the executive director of product management for McKesson Enterprise Intelligence solutions. Her primary focus is identifying and evaluating strategies and solutions that drive intelligence to the healthcare enterprise. In previous roles she served as a healthcare consultant and educator and has published articles for Healthcare Financial Management Association and Hospitals and Heath Networks' Most Wired.

Jason Burke is the former director of health and life sciences at SAS. He coordinated the development and execution of SAS's industry strategy and solutions portfolio across pharmaceutical, healthcare provider, health plan, biotechnology, and regulatory organizations around the world. Jason was the founding director of SAS's Health and Life Sciences Global Practice. He regularly consults with industry leaders and executives on emerging business and technology trends, especially those related to bridging the health and life sciences ecosystems. Prior to joining SAS, he worked at a variety of companies such as GlaxoSmithKline, Quintiles Transnational, and most recently Microsoft where he drove the development and adoption of Microsoft's industry technology strategy and corresponding architecture. He has served as a technology leader in several industry think tank and strategic development initiatives focusing on the future of healthcare and pharmaceutical research.

Kyle Cheek is director of the Center for Research in Information Management in the College of Business Administration at the University of Illinois—Chicago. In that capacity he is responsible for the Center's mission to develop opportunities for faculty and student engagement with the local business community to address practical problems in information management. He also holds an appointment as Clinical Professor of Information and Decision Sciences, and teaches courses on enterprise analytics and healthcare information and analytics.

Prior to his current role, Kyle served in various executive roles in the healthcare industry. His expertise is centered on the application of advanced business analytics to healthcare business domain problems including healthcare informatics, payment integrity analytics, and healthcare information management. Some of his notable accomplishments include leading the definition, development, and implementation of an award-winning advanced-analytic healthcare fraud detection solution, and leading the development and execution of an advanced healthcare analytics business strategy. He is active with numerous professional organizations, both in the healthcare industry and more broadly focused on the furtherance of analytics as a business asset, and is frequently invited to provide commentary to various industry and media forums on the broader adoption of advanced analytics across the healthcare domain.

Kyle received a PhD in Political Economy from The University of Texas at Dallas in 1996.

Thomas H. Davenport is the cofounder and research director of the International Institute for Analytics, and the President's Distinguished Professor of Information Technology and Management at

Babson College. He is the coauthor of *Competing on Analytics* and *Analytics at Work*.

Dave Dimond is the National Leader of Healthcare Consulting at EMC. He is a strategic business advisor and healthcare management consulting leader with more than 20 years of professional experience. Dave received his Master of Science in Management Engineering from Northeastern University, Graduate School of Engineering and his B.S. in Electrical Engineering with High Distinction from the University of Massachusetts College of Engineering. Dave is a candidate in the Executive Scholars Program, Operations Research and Supply Chain Management at Kellogg University.

Dan Gaines is the Provider Health Analytics Solution Lead in Accenture's Practice where he oversees the design and development of next generation health analytic solutions. Prior to joining Accenture, Dan ran the BI and Analytics product engineering organization at the Advisory Board Company, where he and his team were responsible for the creation and evolution of the company's analytic Software as a Service offerings for hospital clients. In addition, Dan's background includes more than 20 years of information management and analytics consulting and product development focused in all aspects of healthcare and life sciences.

Glenn Gutwillig is an executive director in Accenture Analytics, where he serves as the global analytics lead for Health and Public Service industry sectors. Glenn also serves as a member of the faculty for Health Analytics at the International Institute of Analytics (IIA) and is a frequent speaker/instructor at both Analytics and Business Intelligence Conferences and related industry events.

Dave Handelsman is a senior industry consultant in SAS's Health and Life Sciences division. His primary focus is identifying those market opportunities where advanced analytics brings dramatic innovations and improvements to the business and science of the health and life sciences industries. In previous roles within SAS, Handelsman served as the principal product manager for Clinical R&D, where he was responsible for guiding the development and market success of SAS's flagship pharmaceutical solution SAS Drug Development.

Evon Holladay, Vice President of Business Intelligence for Catholic Health Initiatives, leads CHI's efforts to provide information to customers and stakeholders to support timely strategic and operational decision making. She is a seasoned professional in architecting BI solutions through running BI operations and has built scalable, high-value solutions for healthcare, telecommunications, retail, and manufacturing. Her focus is on creating a cross-functional collaborative whereby the information is utilized and maintained. Evon has a passion for designing an operating solution that provides measurable value. Her specific areas of interest are enterprise information integration, data quality management, and working with business partners to build information solutions that improve the quality of care delivered.

Stephan Kudyba is founder of the analytic solutions company, NullSigma, and is also a faculty member of the management department at the New Jersey Institute of Technology, where he teaches courses that address the utilization of IT, advanced quantitative methods, business intelligence, and information and knowledge management to enhance organizational efficiency. He has published numerous books, journal articles, and magazine articles on strategic utilization of data, IT, and analytics to enhance organizational and macro productivity. His latest book, *Healthcare Informatics: Achieving Efficiency and Productivity* (foreword by Jim Goodnight), addresses the critical topic of leveraging new technologies, data, and analytics to achieve efficiency in healthcare.

Dwight McNeill, Ph.D., MPH, is a Lecturer at Suffolk University Sawyer Business School, where he teaches population health and health policy. He is President of WayPoint Health Analytics, which provides guidance to organizations on the analytics of population health management, behavior change, and innovation diffusion. He is the author of *A Framework for Applying Analytics in Healthcare: What Can Be Learned from the Best Practices in Retail, Banking, Politics, and Sports* (FT Press 2013) and numerous journal articles including "The Value of Building Sustainable Health Care Systems: Capturing the Benefits of Health Plan Transformation" (Health Affairs).

Over his thirty year career in healthcare, he has held analytics positions in corporations at IBM and GTE; governments at the Agency for Healthcare Research and Quality and the Commonwealth of Massachusetts; analytics companies; and provider settings.

Jeffrey D. Miller is vice president and leads the Health and Life Sciences business in North America for Capgemini. He oversees the sales and service delivery functions for this industry, driving overall effectiveness and impact across the Capgemini portfolio.

Robert Morison has been leading business research in professional services firms for more than 20 years and is the coauthor of *Analytics at Work: Smarter Decisions, Better Results* and *Workforce Crisis* (both Harvard Business Press). Mr. Morison is also a faculty member with IIA.

Thad Perry has more than 20 years of healthcare experience and is currently the vice president and general manager responsible for all health plan clients within the healthcare business of Thomson Reuters. Prior to 2011, he was director of healthcare informatics at CareSource, chairman and CEO of Health Research Insights, Inc., vice president of healthcare informatics at American Healthways, and held various positions at CIGNA. Dr. Perry earned his M.A. and Ph.D. in Psychology from Vanderbilt University and founded the Medicare and Medicaid Contractors Statistical and Data Analysis Conference, which is now in its 16th year.

Pat Saporito's role at SAP is to work closely with current and prospective customers to leverage best practices in business intelligence, their data assets, and SAP Business Objects business analytic solutions to improve business performance. Pat joined SAP Business Objects in 2006 as insurance solutions director within the Enterprise Performance COE. She subsequently has held roles in global consulting helping customers define effective performance analytic solutions and in solution management developing predefined industry applications. She is a recognized insurance and healthcare payer industry analytics thought leader.

Dean Sittig is a professor at the School of Biomedical Informatics at the University of Texas Health Science Center at Houston. His research interests center on the design, development, implementation, and evaluation of all aspects of clinical information systems. In addition to measuring the impact of clinical information systems on a large scale, he is working to improve our understanding of both the factors that lead to success, as well as the unintended consequences associated with computer-based clinical decision support and provider order entry systems. To this end, he and Hardeep Singh, MD, have proposed a new eight-dimension, sociotechnical model for safe and effective use of health IT.

David Wiggin is the Program Director, Healthcare and Life Sciences, for Teradata Corporation. His responsibilities include industry strategy, marketing, offer development, and field enablement and support. Prior to joining Teradata, David was with Thomson Reuters for 25 years supporting employer, health plan, provider, and government markets. He has worked in a variety of roles, including product management, product development, project management, data warehousing, operations management, and systems architecture. David has experience with both the business and IT dimensions of the healthcare industry and has held executive and management positions at Stern Stewart & Co. and Exxon Corporation.

Jesus Zarate, Col, USAF, MSC; Air Force Medical Service CIO.

Introduction

Dwight McNeill

There has never been a better window of opportunity for analytics to strut its stuff and contribute to dramatic improvements in clinical and business outcomes in the healthcare industry.

- The opportunities are mind-boggling. Clinical outcomes are the worst when compared to peer, wealthy countries. Efficiency is the worst among all industries with at least a third of the healthcare industry's output considered waste. The likelihood of getting the right care at the right time remains just above the probability of a coin toss. Customer engagement ranks among the lowest of all industries.
- The drivers for change are strong and convergent. These include sweeping changes in the financing, payment, and delivery of healthcare resulting from the Affordable Care Act as well as from hypercompetitive market pressures to markedly reduce costs, increase market share, and increase revenues.
- The analytic workbench is chock full of statistical tools, methods, and theories to collect, organize, and understand data and to influence decision making.
- The explosion of "big" data and the technology to harness it more quickly and cheaply provide greenfield opportunities for new discoveries and applications, such as genomics.

Yet, despite these convergent forces, the funding and utility of analytics in healthcare have been low. The irony is that healthcare is built on strong analytic pillars in its extensive research on the causes and treatments of diseases, but this culture and expertise have not spilled over into the delivery of care. Indeed only a dozen or so of the best providers and payers approach the full optimization of analytics.

There are many rationales for this. Among them are

- Improving clinical outcomes and efficiency does not necessarily make good business sense. After all, the healthcare industry is profitable, and the pursuit of social (health) goals is not always aligned with the pursuit of profits.[1] If the industry does not want to change, there is little call for the analytics to support it.

- There are strong beliefs that the industry data are under-digitized, and business cannot benefit from analytics until the data are complete, clean, and perfectly integrated.

- Medical care is delivered by highly trained, autonomous, and intuitive-thinking professionals (doctors) who may eschew data-driven decision making.

- Technology is a two-edged sword. On the one hand it offers awesome capabilities to process data. On the other hand, it may blind analysts from seeing all that is necessary to make change happen through analytics for their industry. The field needs to look inward and transform itself to be more results oriented.

- Finally, the (new and improved) discipline of analytics is relatively young, unknown, and yet to prove itself.

The primary purpose of this book is to address the last bullet point. The book provides the most comprehensive review of the current state of the science and practice of analytics in healthcare to date. The book is divided into a journey of five parts, the four Parts and the Conclusion. For a simple guide on navigating the book see Figure I.1.

Part I, "An Overview of Analytics in Healthcare and Life Sciences," provides an overview of the analytics landscape in the healthcare and life sciences ecosystem and includes chapters on payers, providers, and life science companies. Tom Davenport and Marica Testa, in Chapter 1, "An Overview of Provider, Payer, and Life Sciences Analytics," conclude that despite the many obstacles, "healthcare organizations have little choice but to embrace analytics. Their extensive use is the only way patients will receive effective care at an affordable cost." Although the maturity level of analytics is low across the ecosystem, many opportunities are outlined in the chapters. For providers, these include meaningful use, accountable care, regulatory compliance, clinical decision support, and more. For payers, these include actively improving the health of their members to be more competitive in the new era of the business to consumer model. And for life science companies, the focus is on research discovery, clinical trials, manufacturing, and sales and marketing. Increasingly the focus will be on personalized medicine to tailor individual treatment programs and on cost-effectiveness analysis to determine the value of therapeutics.

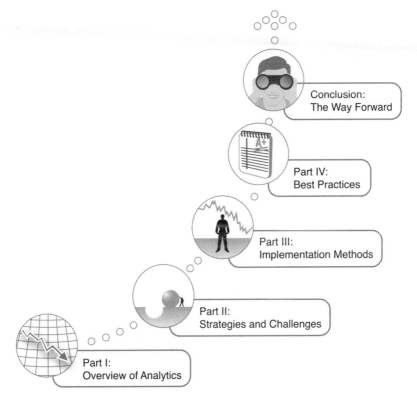

Figure I.1 Navigating the five parts of *Analytics in Healthcare and the Life Sciences*

Part II, "Strategies, Frameworks, and Challenges for Health Analytics," includes six chapters that provide some fundamental answers as well as a reference library of terms and concepts to those wanting to get into the new game of health analytics. It provides a mapping of healthcare analytics "DNA" and addresses the following important questions:

- What is health analytics and what is its scope and various options?
- What is its value to the business and how is it determined?
- What are the different types of analytics and ways to perform them?
- What are some examples of analytics "secret sauce" for supporting clinical and business outcomes?
- What are the privacy concerns that arise with big (personal) data and the use of advanced analytics and how can these be built into data security and privacy practices?

Part III, "Healthcare Analytics Implementation Methods," looks at implementation methods, or solutions. It provides a workbench of analytics methods that address some of the most vexing issues in healthcare, including

- **Using the EHR**—The electronic health records (EHR) can support meaningful results in three important healthcare reform areas, including insurance reforms (especially health insurance exchanges), Centers for Medicare and Medicaid Services (CMS) Innovations (especially Accountable Care Organizations), and Health Information Technology (HIT) (especially meaningful use). Chapter 12, "Meaningful Use and the Role of Analytics: Complying with Regulatory Imperatives," by Deborah Bulger and Kathleen Aller makes the case that building the infrastructure for meaningful use has much more value than just a compliance issue and can be the backbone to a new approach to managing care.
- **Improving the delivery of care**—In Chapter 13, "Advancing Health Provider Clinical Quality Analytics," Glenn Gutwillig and Dan Gaines focus on measuring, monitoring, and improving providers' adherence to established clinical standards. They assert that clinical quality analytics must measure a health provider's compliance to established clinical standards of care as

well as analyze the relationship between compliance and clinical outcomes.

- **Medical errors**—Medical errors continue to be seemingly intractable to improvement. In Chapter 14, "Improving Patient Safety Using Clinical Analytics," Dean Sittig and Stephan Kudyba concentrate on the detection of errors and discuss the use of "triggers," or automated algorithms, to identify abnormal patterns in laboratory test results, clinical workflows, or patient encounters.

- **Social media**—The use of social media to improve health is just emerging. Healthcare is following industries that have used social media for marketing, sentiment analysis, and brand management. Chapter 16, "Measuring the Impact of Social Media in Healthcare," by David Wiggin provides an overview of current and emerging uses of social media to improve health and proposes an analytical model to measure its impact. He suggests that the best source of data may come directly from people through surveys rather than what can be "scraped" from websites. This is different, and potentially much more valuable, than the usual "scraping" of websites for social media data.

- **Population health**—One aspect of managing population health is to find high cost/clinical need people so that appropriately tailored programs can be offered to them. In Chapter 15, "Using Advanced Analytics to Take Action for Health Plan Members' Health," Kudyba, Perry, and Azzolini detail the difficulty of developing, implementing, and managing population-based care programs. They present a conceptual framework, based on "hot spotting" techniques, that defines the information requirements, analyses, and reporting that will lead to actionable results.

Part IV, "Best Practices in Healthcare Analytics Across the Ecosystem," includes eight case studies of leading organizations in healthcare analytics. These are bellwether organizations that represent the best of the art and science of analytics as of 2012. The case studies are inclusive of the settings where analytics is practiced including providers, payers, and a life sciences company and includes both the public and private sectors. The lineup includes Partners HealthCare System,

Catholic Healthcare Initiatives, Veterans Health Administration, Air Force Medical Services, HealthEast Care System, Aetna, EMC, and Merck. The chapters address the "whats" (the domains of the content such as business, clinical, and marketing) and "hows" of analytics to support organizational strategies and goals (including how it is organized, how it adds value, and its technical challenges). The common characteristics of these high performing companies are the early adoption and use of EHRs, leadership that clearly articulates organizational mission and goals, the use of clinical warehouses to address organizational needs such as research, the application of analytics to improve business and finance functions, and insights into how to operationalize analytics within organizations for optimal results.

Finally, a conclusion, "Healthcare Analytics: The Way Forward," addresses the future of analytics in healthcare. It starts off by acknowledging that healthcare has great challenges and that the potential of analytics to address them has been underrated. Analytics is poised to make a difference, but there is a blockage that must be addressed. The conclusion addresses some untapped opportunities including issues related to the big data "gold rush" and the need to appreciate "small" data, new computing technologies such as NoSCL and the seductive trap of technology, and the overlooked science of making change happen and getting innovations adopted in organizations. McNeill suggests that analysts must keep their eyes on the prize of improving outcomes, and it has less to do with the tools and technology and more to do with the sociology of making change happen through communications among people. He concludes with the observation that the field of analytics is undergoing an identity crisis, the role definition needs work, a Chief of Analytics may not be the savior, and what is needed is for analysts to "be the change" they want to see in the organization and the world.

Note

1. Eduardo Porter, "Healthcare and Profits, a Poor Mix," *The New York Times*, January 8, 2013.

Part I
An Overview of Analytics in Healthcare and Life Sciences

1

An Overview of Provider, Payer, and Life Sciences Analytics

Thomas H. Davenport and Marcia A. Testa

The healthcare industry is being transformed continually by the biological and medical sciences, which hold considerable potential to drive change and improve health outcomes. However, healthcare in industrialized economies is now poised on the edge of an analytics-driven transformation. The field of analytics involves "the extensive use of data, statistical and quantitative analysis, explanatory and predictive models, and fact-based management to drive decisions and actions."[1] Analytics often uses historical data to model future trends, to evaluate decisions, and to measure performance to improve business processes and outcomes. Powerful analytical tools for changing healthcare include data, statistical methods and analyses, and rigorous, quantitative approaches to decision making about patients and their care. These analytical tools are at the heart of "evidence-based medicine."

Analytics promises not only to aid healthcare providers in offering better care, but also more cost-effective healthcare. Several textbooks have been written on the cost-effectiveness of health and medicine, and health economics and the methods described can be used in healthcare decision making.[2, 3, 4] Moreover, as healthcare spending rose dramatically during the 1970s and 1980s in the United States, an increased focus on "market-driven" healthcare developed.[5] Today, as the amount spent on healthcare has risen to nearly 20% of GDP in the United States, analytic techniques can be used to direct limited resources to areas where they can provide the greatest improvement in health outcomes.

Analytics in healthcare is an issue for several sectors of the healthcare industry involving patients, providers, payers, and the healthcare technology industries (see Figure 1.1). As shown, the patient is the ultimate consumer within the healthcare system. This system consists of several sectors, including providers of care; entities such as employers and government that contribute through subsidized health insurance; and life science industries, such as pharmaceutical and medical device companies.

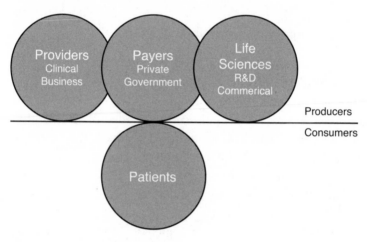

Figure 1.1 The healthcare analytics environment

Provider Analytics

A key domain for the application of analytics is in healthcare provider organizations—hospitals, group practices, and individual physicians' offices. Analytics is not yet widely used in this context, but a new data foundation for analytics is being laid with widespread investments—and government subsidies—in electronic medical records and health outcomes data. As data about patients and their care proliferate, it will soon become feasible to determine which treatments are most cost-effective, and which providers do best at offering them. However, to maximize their usefulness, analytics will have to be employed in provider organizations for both clinical and business purposes and to understand the relationships between them.

Tom Davenport and Jeffrey Miller in Chapter 2, "An Overview of Analytics in Healthcare Providers," make the case that analytics for healthcare providers is poised to take off with the widespread digitization of the sector. They describe the current maturity level of provider analytics as low and describe current analytical applications along the continuum of descriptive, predictive, and prescriptive for both clinical and financial business purposes. And they address future areas for analytics contributions including meaningful use, accountable care organizations, taming the complexity of the clinical domain, increased regulatory requirements, and patient information privacy issues.

Payer Analytics

Payers for healthcare, including both governments and private health insurance firms, have had access to structured data in the form of claims databases. These are more amenable to analysis than the data collected by providers, who have relied largely on unstructured medical chart records. However, historically payers focused on collecting data that ensure efficiencies in billing and accounting, rather than healthcare processes and outcomes. Even with limited administrative databases, payers have, at times, been able to establish that some treatments are more effective and cost-effective than others, and these insights have sometimes led to changes in payment structures. Payers are now beginning to make inroads into analytics-based disease management by redesigning their information databases to include electronic medical records. However, there is much more to be done in developing medical information databases and systems and employing analyses within payer organizations. In addition, at some point, payers are likely to have to share their results with providers, and even patients, if systemic behavior change is to result.

Kyle Cheek in Chapter 3, "An Overview of Analytics in Healthcare Payers," concentrates on analytics as a value driver to improve the business of health insurance and the health of its members. He provides a framework of the types of analytics that can add value, and he reviews the current state, which he describes as "analytical sycophancy." He concludes with paths to maturity and best practice examples from leading organizations.

Life Sciences Analytics

Life sciences companies, which provide the drugs and medical devices that have dramatically changed healthcare over the past several decades, have also employed analytics much more than providers. However, their analytical environment is also changing dramatically. On the R&D and clinical side, analytics will be reshaped by the advent of personalized medicine—the rise of treatments tailored to individual patient genomes, proteomes, and metabolic attributes. This is an enormous (and expensive) analytical challenge that no drug company has yet mastered. On the commercial analytics side, there is new data as well—from marketing drugs directly to consumers, rather than through physicians—and new urgency to rein in costs by increasing marketing and sales effectiveness.

Dave Handelson in Chapter 4, "Surveying the Analytical Landscape in Life Sciences Organizations," starts off with the contextual reality that it is no longer "business as usual" in the life sciences industries, which has resulted in a heightened focus on analytics. He describes the potential analytical contributions related to the primary business functions, including research discovery, clinical trials, manufacturing, and sales and marketing. He notes that healthcare reform and the emphasis on cost containment place more reliance on analytics that includes new reimbursement strategies and the need to use comparative effectiveness results in assessing the value of therapies.

Patients Analytics

Patients are, of course, the ultimate consumers of healthcare and will need also to become better informed consumers of analytics—at least to some degree. They will need analytics to decide which providers are most effective, whether the chosen treatment will work, and in some payment structures, whether they are getting the best price possible. These consumer roles are consistent with the "consumer health informatics" and "Health 2.0" (use of web-based and e-technology tools by patients and physicians to promote healthcare and education) concepts. Of course, complex biostatistics and the results of comparative effectiveness studies are unlikely to be understood by most patients and will have to be simplified to be helpful.

Collaboration Across Sectors

Each of the sectors that participate in healthcare progressively adds analytical capability, although at different rates. For true progress, analytics must be employed collaboratively across the various sectors of the healthcare system. Providers, payers, and pharmaceutical firms must share data and analyses on patients, protocols, and pricing—with each other and with patients—and all with data security and privacy. For example, members of each sector had data that might have identified much earlier that COX-2 drugs (Vioxx, Celebrex, and Bextra) were potentially associated with greater risk of heart disease.

Barriers to Analytics

Healthcare organizations desiring to gain more analytical expertise face a variety of challenges. Providers—other than the wealthiest academic medical centers—have historically lacked the data, money, and skilled people for analytical projects and models. Even when they are able to implement such systems, they may face difficulties integrating analytics into daily clinical practice and objections from clinical personnel in using analytical decision-making approaches. Payers typically have more data than providers or patients, but as noted above the data are related to processes and payments (administrative databases) rather than health outcomes (research databases). Moreover, many payers do not now have cultures and processes that employ analytical decision making.

Life sciences firms have long had analytical cultures at the core of their research and clinical processes, but this doesn't ensure their ongoing business success. Clinical trials are becoming increasingly complex and clinical research more difficult to undertake given the restrictions imposed by Institutional Review Boards, ethics committees, and liability concerns. Drug development partnerships make analytics an interorganizational issue. And the decline of margins in an increasingly strained industry makes it more difficult to afford extensive analytics.

While statistical analyses have been used in research, analytics has not historically been core to the commercial side of life sciences industries, particularly in the relationship with physicians' practice patterns. Life sciences firms must normally buy physician prescribing data from a third-party source, and the data typically arrive in standard tables and reports rather than in formats suitable for further analysis. The firms increasingly need to target particular physicians, provider institutions, and buying groups, but most do not have the data or information to do so effectively.

Despite these obstacles, healthcare organizations have little choice but to embrace analytics. Their extensive use is the only way patients will receive effective care at an affordable cost.

Notes

1. Thomas H. Davenport, Jeanne G. Harris, *Competing on Analytics* (Boston, MA: Harvard Business Press, 2007) p. 7.

2. Marthe R. Gold, Joanna E. Siegel, Louise B. Russell, and Milton C. Weinstein, *Cost-Effectiveness in Health and Medicine* (New York, NY: Oxford University Press, 1996).

3. Rexford E. Santerre and Stephen P. Neun, *Health Economics: Theories, Insights, and Industry Studies*, 5th Edition (Mason, OH: South-Western Cengage Learning, 2010).

4. Sherman Folland, Allen Goodman, and Miron Stano, *The Economics of Health and Healthcare*, 6th Edition (Upper Saddle River, NJ: Prentice Hall, 2009).

5. Regina E. Herzlinger, *Market Driven Healthcare: Who Wins, Who Loses in the Transformation of America's Largest Service Industry* (Cambridge, MA: Basic Books, A Member of the Perseus Books Group, 1997).

2

An Overview of Analytics in Healthcare Providers

Thomas H. Davenport and Jeffrey D. Miller

For the most part, analytics is just beginning to be employed in most healthcare provider organizations. However, it offers a high degree of potential to immediately improve patient safety and the financial and operational performance of the healthcare enterprise. Given the rapid rise in adoption of electronic medical records (and reimbursement based on their "meaningful use"), we expect to see considerable advances in the use of analytics by providers over the next five years as analytics provide the foundation for improved clinical quality and reduced cost of care.

There are two major areas of analytical activity in provider organizations:

- **Financial and operational**—These areas include analytical applications to monitor, predict, and optimize facilities, staffing, admissions, reimbursements, and other key factors driving the performance of a provider institution.
- **Clinical and patient safety**—This domain includes analytics related to evidence-based medicine and clinical decision support, comparative effectiveness, patient safety, survival rates, and compliance with care protocols.

In both areas, descriptive analytics (reporting, scorecards, dashboards, and so on) have been the primary focus (as compared to predictive analytics or prescriptive analytics). In many cases, reporting has been driven by increased regulatory requirements or narrowly defined performance improvement programs (e.g., reduced supply costs). Predictive analytics have largely been employed for forecasting patient admissions (in some cases by service line) on the business and operational side, and for scoring patients likely to need intervention on the clinical side—but largely in academic medical centers and on a pilot basis.

Overall, analytics at providers are highly localized, generally within department or functional boundaries. More sophisticated providers have pockets of analytical activity in various places, but little coordination. One leading academic medical center, with a strong history of clinical decision support and some degree of business analytics as well, had never had any contact between the two groups until a recent task force on "meaningful use."

In terms of the analytical capability framework in Analytics at Work (see Figure 2.1), few provider institutions go beyond Stage 3, "Analytical Aspirations." Most community hospitals are probably at Stage 1 ("Analytically Impaired"), because they lack electronic medical record (EMR) data (as of 3Q 2010, only 50% of hospitals with fewer than 200 beds had achieved the third of the seven stages of EMR adoption), and because they lack the human and technology resources to analyze financial and operational data. Mission-focused (religious or charitable) providers may find it particularly difficult to marshal the resources to perform serious analytics given the many demands on the limited capital budgets of these institutions.

The larger for-profit hospital chains and integrated delivery networks (IDNs) are largely at Stage 2 ("Localized Analytics"). Because of their increased focus on financial and operational performance, analytics are used primarily on the business side in this type of provider. However, the rise of EMRs and pressure for cost-effective treatments will create a more balanced clinical/business analytics environment over the next several years.

Academic medical centers are also typically at Stage 2, although their analytical focus is more likely to be on the clinical side. By virtue of their missions they are more likely to do clinical research, comparative effectiveness studies, clinical decision support implementations, and so forth. Many have chief medical information officers or medical directors with a focus on clinical decision support. Some may have a degree of descriptive (and some predictive) analytics on the business side as well. However, the analytics remain largely localized, and for the great majority have not been viewed as an enterprise resource.

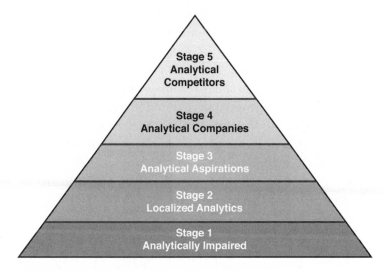

Figure 2.1 Five levels of analytical capability

Only a few healthcare organizations are at Stage 3, "Analytical Aspirations." Those that qualify are primarily integrated care organizations (Kaiser Permanente, the Mayo and Cleveland Clinics) and a few academic medical centers (Intermountain, Partners, Johns Hopkins). These organizations are still largely localized in their analytical focus, but they have several distinguishing characteristics:

- Their localized analytics groups are beginning to communicate and collaborate.
- They have high degrees of analytical activity and personnel (particularly on the clinical side), some of which are world-class.
- Their senior management teams recognize that analytical capabilities are key to their success.

However, because of organizational boundaries, resource limitations, disparate technologies, and other constraints, they cannot yet be classified as "analytical organizations" or "analytical competitors" compared to standard-bearers for those titles in other industries.

Analytical Applications in Healthcare Providers

In this section we describe the key analytical applications used in provider organizations and classify them as to whether they involve descriptive or predictive/prescriptive analytics (see Table 2.1). The latter types use past data to construct predictive models (predictive analytics), or use past data to suggest action (prescriptive analytics), such as the most effective treatment to employ, or the optimum level of staffing to use. Where appropriate, we describe each application as to how granular it typically is in terms of data. As more granular levels of analysis are possible—for example, moving from institutional data to service line data to procedure data—analytics are easier to translate into action. Highly granular data may also help lead to behavior change; for example, reporting certain data at the individual clinician level tends to bring out a sense of competition and a desire to excel against peers.

The Future of Analytics in Provider Organizations

As we have suggested, the future of analytics in healthcare provider organizations is much brighter than in the past. A number of factors will drive considerably greater activity in the future. Several are described in this section.

The "meaningful use" criteria for EMR reimbursement are clearly going to lead to more analytical activity.[1] This is one of the first times in any industry that reimbursement for a system demands actually using the data contained in the system! The criterion involving clinical decision support rules, for example, presumes some analytical underpinnings. Others involve specific reports, such as, "Generate lists of patients by specific conditions to use for quality improvement, reduction of disparities, research, and outreach." While most of the 25 Stage 1 meaningful use criteria are more transactional in nature than analytical, future meaningful use criteria are likely to be more analytically focused.

Table 2.1 Key Analytical Applications in Provider Organizations

	Descriptive Analytics	Predictive Analytics
Financial/ Operational	Occupancy analytics—Models reporting on or (much less common) forecasting and predicting bed occupancy, either across the institution or within particular service lines	Population analytics—Predictive models (usually provided by third parties, and often based on Medicare claims data) of population demand in a particular service area used for physician and facility planning, either for overall population health or by disease/service line
	Revenue cycle analytics—Reporting on billing, collections, and accounts receivable, most frequently at the institutional level, but occasionally at the payer or service line level	Financial risk—Predictive models of patients' likelihood of self-payment or reimbursement, usually at admission; granular to the patient level
	Quality and compliance analytics—Reporting models assessing compliance with regulations and payer requirements, usually at the institution level	Supply chain analytics—Reporting on, and in some cases, optimizing levels of inventory for hospital supplies—sometimes in partnership with an external supply company or group purchasing organization (GPO)
	Cost management—Reporting models of staff productivity and cost, typically by service line, and comparisons to industry benchmarks	
	Operational benchmarks—Reporting on basic measures of operational performance, such as length of stay, percentage discharged by particular time—usually at institutional level, but increasingly at service line and even individual physician level	

	Descriptive Analytics	Predictive Analytics
Clinical	Clinical problem tracking—Reporting and tracing particular clinical problems, such as post-operative infections, which caregivers touched the patient, which other patients were touched by those caregivers, and so on	Prediction of likely treatment results—Based on attributes of the patient's biodata, symptoms, and metabolic data, analytical predictions of the likelihood of acquiring certain diseases and of successful treatment can already be made, although they primarily are done only in experimental mode by academic medical centers at this time
	Patient safety analytics—Reporting on common safety problems, such as central line infections, generally at entire hospital level	Scoring of patients at risk—Using analytics to identify predictive variables for risk of particular problems (e.g., cardiovascular failure), and then scoring patients to identify those most at risk; presently primarily in pilot mode
		Care protocol analytics—Controlled research studies determining which treatment protocols are most effective; usually only in academic medical centers
		Evidence-based medicine analytics—Translation of clinical research results (at the particular hospital or elsewhere) into order sets and rules for provider order entry (CPOE) systems; may be incorporated into EMR and CPOE systems from external vendors
		Personalized genetic medicine—Predictions of medication choice and dosing based on genetic profile, presently in pilot mode at some academic medical centers

The move to accountable care organizations (ACOs) will clearly drive providers toward greater reliance on descriptive analytics. Since reimbursement for physicians and hospitals will depend on their ability to control costs and meet quality-of-care goals, metrics of those factors will be highly prized. It is likely that providers will move toward highly granular reports—down to the individual clinician level—to motivate the desired (and reimbursed) behaviors. ACOs will need to blend financial, operational, and clinical analytics in a way that they have rarely been integrated in the past.

Analytics will also be required to address *growing complexity in the clinical domain.* There is arguably already too much medical knowledge for humans to hold in their brains, so clinical decision support based on analytics will be necessary to deliver the best care. Personalized genetic medicine presents an even more daunting amount of content to master. Providers will need to transform the roles of their clinicians toward greater use of analytics and decision support, as opposed to relying primarily on experience and intuition.

Analytics will also be advanced through *increased regulatory requirements for reporting and transparency.* There are already substantial institutional compliance issues from organizations like Centers for Medicare and Medicaid Services (CMS), Food and Drug Administration (FDA), and so on. As more data at the patient and clinician levels become available, CMS and other regulatory bodies will undoubtedly require more data. Caregivers at every level will be assessed on whether they followed a prescribed protocol, whether they have documentation for protocol deviations, and so forth.

Patient information privacy issues won't necessarily advance analytics, but they will certainly influence them. This has not yet been an issue at the aggregated data level, but with more atomic data, privacy issues will be much more difficult to address. It will also make outsourcing of analytical services more challenging. Questions about the use of patient data for analytical purposes abound. For example, if an institution sends out a patient profile for cancer screening, can the organization receiving it guarantee that it will remain private? Will patients have to opt in to sharing data for analytical purposes, even when it has been anonymized? Sharing data across payer and provider networks may be impossible without the patient's permission to use the data.[2]

In short, the future analytical world in which healthcare providers will live will be very different from that of today. Many of the analytical initiatives and programs started in the current environment will have to be exploratory, but they will undoubtedly confer both experience and advantage on early adopters. Uncertainty is no excuse for inaction.

Notes

1. For a concise list of meaningful use objectives, see the *Healthcare IT News* list at http://www.healthcareitnews.com/news/eligible-provider-meaningful-use-criteria.

2. For additional reading, see "Security, Privacy and Risk Analytics in Healthcare" published in the HLSARC compendium.

3

An Overview of Analytics in Healthcare Payers

Kyle Cheek

For every complex problem there is a solution that is simple, neat, and wrong.

—H. L. Mencken

Quantitative analytics have long been utilized by healthcare payer organizations, tracing back to the actuarial origins on which the payer industry is based. In spite of its quantitative origins, though, the deployment of analytics as a competitive driver beyond the traditional actuarial pricing function has been slow and uneven, even as healthcare payers have evolved from their traditional indemnity role to one of trusted health advocate—a transition that is inherently information-centric and demands a greater reliance on deeper analytics. Payers do generally recognize that there is latent value in the large volumes of data they accumulate through healthcare transaction processing, especially in the mountains of healthcare claims data that contain rich financial and clinical information and, when properly aggregated, provide rich longitudinal views of patient experiences.

Historically, though, efforts to exploit that value have largely disappointed, either because of a failure to employ the deep analytics that drive competitive advantage or because deep analytics have been adopted without being fit holistically into the program they support. The failure to employ deep analytics is seen in the ubiquitous reporting applications—from employer group reporting applications to executive dashboards—that are driven by simple summary views of the underlying information. That focus on reporting applications and other less sophisticated uses of payer data resources has tended to be accompanied by a concomitant focus on technology as the delivery mechanism, rather than the sophistication of the actual analytics delivered. Similarly, the adoption of more sophisticated analytics in the payer space has often occurred within programs that are not properly designed to avail themselves of deep quantitative results or, conversely, that produce value across the longer time horizon necessary to realize value from efforts to effect changes in patient behavior, but concurrently making efforts to measure that value more difficult.

While the adoption of analytics as a competitive catalyst has met with mixed success industry-wide, there do exist payer organizations that have made a principled commitment to exploit analytically derived value and stand above the industry norm. These organizations typically have fact-driven leadership and a comprehensive strategy for analytics as a value driver, and a commitment to the infrastructure necessary to support an enterprise analytical vision. The successes these organizations have experienced and the means by which they have realized those successes serve as a bellwether for the industry to follow toward more effective adoption and deployment of analytics as a competitive driver and as a real source of change in patient outcomes and population wellness.

Payer Analytics: Current State

Adoption

That healthcare payers have struggled to realize significant value from analytics does not mean that there have not been important forays into analytics within the payer space. Rather, payers have made a concerted effort to leverage their formidable data assets in several distinct areas. Examples of those solutions by analytical subject include

- **Patient analytics**—Many payers have adopted solutions that are focused on the analysis of patient outcomes as a means of predicting and reducing avoidable costs. These disease management (DM) tools typically focus on identifying those patients at greatest risk of poor (and costly) health outcomes and prioritizing clinical intervention to those whose predicted costs are most actionable (often by a nurse-administered telephone outreach call or by direct mail).

- **Provider analytics**—Another class of analytic solutions employed by payers to promote improved outcomes and cost containment includes evidence-based medicine (EBM) and pay for performance (P4P). These analytically based programs focus on provider efficacy and rely on empirical analytics to determine which treatments provide superior outcomes at lower cost. In turn, providers are encouraged to employ more cost-effective treatments, with reimbursement levels depending on the extent to which a provider complies with EBM standards. Similarly, provider transparency programs promote informed patient choice by helping patient populations assess which providers are most compliant with EBM standards and whose patients experience the best health outcomes. All of these analyses require the extraction of deeper insights from the data than simple reporting or basic analyses can provide.

- **Customer analytics**—Employer group reporting and directed marketing are examples of payer efforts to utilize information assets to better allow the purchaser of benefits to manage costs. In addition, some payers have even adopted analytics heavily utilized in direct marketing to address attrition and churn, especially within the individual products segment.

- **Financial/operations analytics**—The final class of analytics adopted by payers is largely inward-facing, and comprises that class of analytics commonly referred to as "business analytics." Solutions widely adopted by payers within this category include financial forecasting, actuarial rating, operations monitoring, and fraud detection.

As the continued proliferation of the preceding examples illustrates, there is awareness among payers of the potential value of analytics. The maturity ascribed to the payer segment must be tempered, though, by a consideration of the effectiveness of these analytic endeavors as a competitive differentiator. Put differently, the general trend among payers to champion analytical solutions has not translated into a corresponding record of clear added value. In one conspicuous example, payers have widely adopted disease management (DM) solutions that rely on analytics to identify patient segments likely to incur significant future costs. Those patients identified by the analyses are then targeted for some sort of intervention—ranging from preventive care service reminders distributed by mail to telephone engagement by clinicians to coach behaviors that promote wellness—designed to reduce overall patient costs. However, in spite of wide adoption and a highly competitive vendor space, there is a body of literature from both public[1] and private payers that maintains there is little empirical evidence that these programs deliver a significant return. Instead, it is argued that payers view these programs as "table stakes"—a cost that must be borne to compete for certain attractive customer segments (typically large group customers).

Another example of widespread analytics adoption is the trend among payers to leverage analytics to combat the problem of fraudulent claims and billing schemes. Solutions that leverage analytics (of varying levels of sophistication) to identify fraudulent claims, providers, and patients have become commonplace in payers' Special Investigation Units. Despite the availability (and large-scale adoption) of tools designed to support the detection of fraud, most authorities still estimate losses to fraud, waste, and abuse at greater than 10% of claims costs. An accepted rule of thumb is that only 10% of those losses are detected through current fraud detection capabilities, and only 10% of those detected dollars are recoverable, suggesting that only about 1% of the total fraud problem is addressed through current antifraud programs and their supporting analytics.

These examples illustrate what might be described as an analytical sycophancy across much of the payer space—a clear recognition that the market demands analytical value and even a demonstrated willingness to adopt, but a marked failure to deploy analytics to meaningfully drive value. Specific to the previous examples, the failure to realize value from disease management programs is often attributed not to shoddy analytics, but rather a failure to build programs that effectively intervene with those patients identified by analytics. In the case of fraudulent claims, the failure to significantly reduce losses was once regarded as a function of inferior analytics, but now is seen more as a product of the deployment of fraud analytics retrospectively—after payments have already been made. In short, any evaluation of the analytical maturity of the payer space should recognize payers' efforts to adopt value-enhancing analytics. However, it is also important to recognize that many such efforts have failed to generate clear-cut successes. In turn, the driver behind the lack of clear analytical success appears to be attributable less to the analytics themselves and more to the efficacy of the programs and processes that are designed to act on analytical results.

Maturity Model

The Davenport-Harris Analytics Maturity Model articulates five levels of analytical competency to describe organizations' capabilities in this domain. From highest to lowest, the levels of Analytical Capability are

- Stage 5—Analytical Competitors
- Stage 4—Analytical Companies
- Stage 3—Analytical Aspirations
- Stage 2—Localized Analytics
- Stage 1—Analytically Impaired

Most payer organizations are best classified at Stage 2 on this model. They have some localized capabilities, as described previously, but have not yet articulated analytical aspirations in a holistic strategy. They probably also have not developed data warehouses or a common analytical toolset at the enterprise level.

Impediments to Maturity

Health payers offer an interesting case study of the evolution of an industry and the capability of its constituent members to adapt to changing competitive dynamics. Considering the evolution of the healthcare industry, and especially the central role played by payer organizations in that evolution, it is not surprising that payers have experienced an uneven transition toward analytical competency. Payers historically operated as transaction-oriented companies, focused on the timely processing of claim, membership, and network transactions. It is of recent vintage that payers have assumed the more analytically-dependent and information-centric role of trusted advisers for the health and wellness of their policyholders. Delivering that evolving focus on information-centric business processes is complicated, though, by the decades of infrastructure that have been built to facilitate the timely processing of transactions.

The vast amounts of data created and warehoused by payers—and upon which any analytical evolution relies—are also a direct artifact of payers' transaction-centric history. As a result, the data that are most readily available to support forays into outcome and wellness analytics (disease management, provider transparency, etc.) are in fact naturally suited for financial analysis rather than for clinical analysis. The manipulation then required to properly condition claims transaction data for use in clinically-oriented studies is not insignificant and is a component of analytical infrastructure that can be prohibitively resource-intensive for smaller payers that cannot make large investments in their data infrastructure.

Despite the seemingly large amount of transaction data available at an individual payer organization, the amount of data may be too small or too narrowly focused to make analytical generalizations to national populations. This problem is being addressed by the rise of national-scope insurers and databases (e.g., United Healthcare and Ingenix), and collaborations among regional or state-based payers (e.g., Blue Health Intelligence). However, these large-scale data aggregations have only recently been compiled, and most of the analytics on them have yet to be performed.

Conclusion: Paths to Maturity

The healthcare payer industry is perhaps best conceived as an earnest occupant of Stage 2 status in the analytical maturity model referenced previously—solidly committed to localized analytics and beginning to develop sincere analytical aspirations. Fortunately, while the payer space as a whole is only beginning to formulate analytical aspirations, there do exist leaders in the space that point to different pathways to maturity for other payers. Among the leading organizations are

- **Aetna**—Aetna is a true analytic competitor, with a rich tradition of fact-based leadership and deep analytics embedded across business processes and products. The Aetna model is best thought of as bottom-up, tracing its origins to localized analytics that grew organically over time to eventually support a wide array of business operations.

- **Humana**—Whereas Aetna's analytics evolution was bottom-up, Humana's commitment to analytics has been top-down. Humana's experience reflects a concerted effort by its leadership to commit to analytics as a value driver. Its analytics organization was created specifically to nurture an enterprise analytics competency that was not previously recognized within the organization.

- **UnitedHealthcare**—United, primarily through its Ingenix subsidiary, has both acquired and developed an impressive array of analytical capabilities. It is one of the few payer organizations with analytical groups serving not only its own needs, but those of other payers and providers as well.

While it requires a major strategic commitment to resources, process, and often even culture change for a payer organization to embark on a path toward analytical excellence, the prospects are good for this segment of the healthcare industry to succeed in this endeavor over time. Significant potential value opportunities have already been identified, for example, in the disease management and fraud detection spaces. The regulatory environment is also supportive of the necessary infrastructure investments and, in general, of a more effective use of analytics to promote better outcomes and efficiencies.

Note

1. Nancy McCall, Jerry Cromwell, Shulamit Bernard, "Evaluation of Phase 1 of Medicare Health Support Pilot Program Under Traditional Fee-for-Service Medicare," Report to Congress by RTI International, June 2007. Online at http://www.cms.hhs.gov/reports/downloads/McCall.pdf.

4

Surveying the Analytical Landscape in Life Sciences Organizations

Dave Handelsman

The life sciences industries are collectively engaged in discovering, developing, and commercializing new therapies. Historically, this market segment has included just the pharmaceutical industry when, in actuality, life sciences organizations encompass pharmaceutical, biotechnology, medical device and vaccine developers, as well as supporting organizations such as contract research (service) organizations and the federal government. Each of these organizations is engaged at overlapping, and multiple, points along the spectrum of bringing new therapies to market, and this chapter puts the analytics landscape in the context of the overall life sciences ecosystem without straying into the nuances of each organizational segment.

For many years, the life sciences industries were viewed as both socially-conscious enterprises and financial powerhouses. These companies developed therapies to treat illness, while at the same time rewarding their shareholders. In recent years, however, this perception is changing due to safety concerns regarding approved therapies, diminishing research pipelines, patent expirations, and a dwindling number of new therapy approvals each year, all of which is coupled with an estimated cost of $1.2B to bring a new therapy to market.[1] Because of these changes, it is no longer "business as usual" in the life sciences industries. This changing business environment, coupled with the added challenges associated with healthcare reform, is forcing life sciences organizations to revisit their decision-making processes. This has led to an increased focus on analytics across the different functions of life sciences firms, although as yet there has been little collaboration or coordination across the functional groups. The most significant analytical functions are reviewed in this chapter.

Discovery

Perhaps no place else in the life sciences industries have analytics been more aggressively pursued than in the discovery process. When the human genome was mapped in 2000, itself a sophisticated analytics project, there was widespread belief that this event would usher in a new collection of cures and treatments. Instead, it ushered in an ever-increasing level of sophisticated analytics, as each new layer of knowledge regarding the fabric of life was studied in complex detail. Microarrays, proteomics, Single-Nucleotide Polymorphisms (SNPs), copy number, and other emerging sciences were studied in detail in an attempt to identify patterns in the spectrum of data that would serve to identify causes or predictors of disease and, ultimately, complementary treatments to address those diseases.

The intensity of analytics associated with this exciting field of research has resulted in not only dedicated analytical software solutions targeted at this market segment, but bespoke hardware strategies to provide the high performance computing required to complete these assessments on a more industrialized basis. The Archon X prize (http://genomics.xprize.org/), like other similar groundbreaking awards, will be awarded to the company that can sequence 100 human genomes, in ten days, at a cost of no more than $10,000 per genome. Parallel efforts are under way to develop advanced analytics to support the $1,000 genome.

While great progress *has* been made in the genomic and associated sciences, it has had limited impact on real-world issues facing the life sciences industries today. Arguably the most success can be seen in the commercialization of Herceptin by Genentech. This treatment, perhaps at the forefront of personalized medicine, has shown great success in treating a specific form of cancer characterized by the *HER2* receptor. This type of analytics-driven result, where a specific diagnostic is paired with a targeted therapy, holds great potential, but has not (yet) been frequently replicated or industrialized despite the sophisticated analytics routinely being applied to the fields of discovery sciences.

Development

For therapies that progress beyond discovery, the development process has routinely employed analytics to determine the safety and efficacy associated with a particular clinical (or preclinical) trial, or an aggregation of the trials that point to the overall safety and efficacy of a new treatment. While the vast majority of the analytics applied in documenting safety and efficacy are descriptive in nature, inferential statistics are used to draw statistically valid conclusions from the available clinical trial data.

The application of these descriptive and inferential analytical approaches is mature and well-understood by both life sciences companies and the various regulatory agencies. The acceptance of this analytical approach by both the manufacturer and the approving agency streamlines the approval process and further supports a mature review process. Additional analytical approaches, such as Bayesian methodology, are emerging as alternative analytical tools in assessing clinical trial outcomes, but these approaches have not yet achieved a similar level of acceptance. In February 2010, the Food and Drug Administration (FDA) issued the *Guidance for the Use of Bayesian Statistics in Medical Device Clinical Trials*, which should further encourage using Bayesian methodology by life sciences organizations. Bayesian approaches, however, are not without controversy, and life sciences companies need to recognize that not all regulatory reviewers are aligned with this emerging analytics approach.

Despite the maturity associated with applying analytics to determine the outcome of a research trial or program, analytics have rarely been applied to other aspects of clinical trial operations. As shown in Figure 4.1, analytics has a strong role to play in the planning and execution of clinical trials. However, this role has typically been limited to less valuable parts of the decision-making process where historical data are summarized (standard reports, query drilldown, etc.), but not used to drive more valuable, optimized, decision-making objectives.

Some analysts have suggested that the absence of clinical trial optimization capabilities, either in human experts or analytical software, has been one factor explaining the relatively limited success that many pharmaceutical firms have had in bringing new therapies to market.

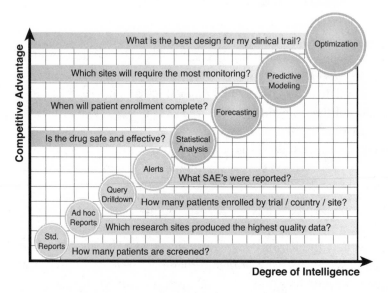

Figure 4.1 Analytical maturity model for clinical trials

When life sciences companies can implement optimized trial designs, as an example, this has a cascading effect throughout the organization. Fewer trials are required to get a therapy approved, fewer resources are required for those trials that are conducted, and the overall risk and expense associated with bringing a new therapy to market are reduced.

Manufacturing

Pharmaceutical manufacturing has been a headline-grabbing issue in recent years as the industry addresses significant product recalls due to manufacturing issues and widespread drug shortages. The problem is sufficiently important that a Drug Shortages Summit was held in late 2010 to address this emerging issue.

These shortages are being caused by a series of related issues including manufacturing difficulties, unexpected interruptions in the supply chain, increased demand, and both voluntary and involuntary recalls. Additionally, some manufacturers have determined that it is simply no longer profitable to produce certain drugs based upon direct costs, revenue opportunity, and risk of litigation.

It is interesting to note that these manufacturing problems are occurring despite the broad use of manufacturing-based analytics for many years. Most notably, Six Sigma methodology has been implemented in the life sciences as one key way to reduce defects in the manufacturing process. Lean manufacturing is another analytics-based discipline focused on reducing waste in the manufacturing process. Hybrid strategies, such as Lean Sigma, focus on reducing both defects and waste, and enable manufacturers to achieve and maintain a high degree of quality in their manufacturing process.

More recently, Quality by Design (QbD) has taken an increasingly important role in life sciences manufacturing. As its name implies, quality is designed into the manufacturing process from the beginning with the goal of ensuring that the issues associated with product and process are carefully considered to ensure quality in the end product.

In May 2007, the FDA issued a report titled *Pharmaceutical Quality for the 21st Century: A Risk-Based Approach Progress Report.* This report specifically identifies QbD as a key strategy for ensuring quality in manufacturing and has shaped the Agency's approach to not only monitoring manufacturing activities but also in paving the way for manufacturers to submit information based on QbD principles. In this same report, the agency outlined the benefit that Process Analytical Technology (PAT) brings to further understand and control the manufacturing process, where *"analytical* in PAT is viewed broadly to include chemical, physical, microbiological, mathematical, and risk analysis conducted in an integrated manner."[2]

As the FDA looks to assess and understand why manufacturing quality continues to be a problem despite the widespread use of analytics, new processes and strategies need to be put in place at the federal level. The number of product recalls by the FDA Center for Drug Evaluation and Research (CDER) continues to grow, and increased by more than 50% in 2009.[3] While it would seem logical for the FDA to increase the number of inspections to identify manufacturing issues more effectively, such an increase is unlikely given the increased costs and the difficult economic climate. The FDA, instead, is turning to a more targeted risk-based approach to identify those manufacturing sites most likely to be at risk for problems.

Sales and Marketing

The sweeping changes affecting the life sciences industries extend far and deep in sales and marketing organizations. Manufacturers are dramatically restructuring their sales teams to address the financial challenges facing the industry, and are no longer deploying a commercialization strategy that relies upon ever-growing armies of sales representatives to sell their products. Instead, life sciences companies are dramatically reducing the size of their sales teams, while at the same time determining how best to "do more with less" in an ongoing struggle to retain market share and grow revenue, despite the changing marketplace.

These changes include the often-mentioned patent cliff, diminishing numbers of new products to sell, and the enormous changes expected to be realized through healthcare reform. Analytics, especially in light of the diminishing number of "feet on the street," will grow in importance for manufacturers. Historically, analytics has meant little more than basic query, rank, and report, where sales analytics were primarily historic reports of past effectiveness, usually sourced from third-party data sources such as IMS Health.

Increasingly, manufacturers are deploying predictive analytics to drive (and, in some cases, recapture) revenue. These predictive analytics span a broad range of the commercial segments, covering such diverse areas as physician targeting, marketing mix, managed care, and rebate optimization. In many ways, the life sciences industries are being forced to adopt mature strategies used in other industries that have not had the luxury of year-over-year revenue growth that has supported their inefficient and expensive sales and marketing practices. Unlike those industries, however, commercial life sciences approaches are more complex, and must carefully consider the world of formularies, reimbursement strategies, and government regulations associated with marketing activities.

In the case of physician targeting, for example, legions of sales representatives would typically be assigned to visit prescribers in their territory that met only the most basic of prescription profiling behaviors—typically those that ranked at some basic level of prescriptions. Instead, predictive physician targeting identifies those physicians that are low-level prescribers but have the potential to not only grow but become high-prescribers with the right level of education from the biopharmaceutical sales team. While the existing high-prescribers can't be ignored, their opportunity for growth is limited, and an intelligent targeting approach can drive increased revenue.

Similarly, new analytical approaches are being applied to more intelligently deploy marketing resources. A wealth of options is available to life sciences manufacturers, and the right mix of media—online, print, radio, and so on—is critical for informing not only consumers, but also physicians and other medical personnel. There is no doubt that direct-to-consumer advertising has an extraordinary impact on driving prescriptions, and it remains increasingly important to be able to monetize marketing investments as the industry gets progressively more competitive.

Unlike other industries, the mechanics of payer reimbursements have a significant effect on purchasing behavior. Life sciences manufacturers frequently structure rebate relationships with providers that share the joint goal of increasing the manufacturers' revenue while reducing the payers' expenses. These relationships are complex and must take into account Medicaid pricing constraints, perceived pricing differences between competitive drugs, and profitability goals. The availability of historic data can be used to construct analytical models that optimize the rebate scenarios for the manufacturer, while taking into consideration other various constraints that must be observed.

Health Reform

The emergence of health reform will weigh heavily on how best to apply new forms of analytics to the business of life sciences development and commercialization. With reform comes new reimbursement strategies, evolving relationships with payers and providers, and the emerging need to more carefully and thoroughly understand the "comparative effectiveness" of new and existing therapies. "Me-too" drugs that provide little-to-no additional therapeutic benefit when compared to older, lower-priced, drugs will present new challenges in terms of driving revenue. Development plans for such drugs are most likely to be deprioritized in favor of differentiated therapies. Such differentiation may be addressed through targeted patient populations, or through renewed focus on novel therapies more likely to both improve patient health as well as the bottom line. In these cases, advanced analytics becomes even more critical, as choices regarding which therapies to pursue and ultimately commercialize will have critical implications regarding the financial viability of life sciences companies. Healthcare reform may also mean that the diverse and siloed analytics groups within life sciences firms will need to collaborate to a greater degree.

Conclusion

Analytics, which have always had a position in the world of the life sciences, are at the forefront of enabling life sciences companies to take their organizations to the next level in today's economy. The issues faced today span discovery through commercialization, and the larger health-care community is struggling to rein in costs while still improving health. Critical scientific and business decisions can only be confidently made by leveraging historical and current data, and bringing advanced analytics to bear on the complex problems facing the life sciences industry today.

Notes

1. Tufts Center for the Study of Drug Development (http://csdd.tufts. edu/).

2. http://www.fda.gov/AboutFDA/CentersOffices/ OfficeofMedicalProductsandTobacco/CDER/ucm128080.htm

3. Parija Kavilanz, "Drug Recalls Surge," CNNMoney.com, August 16, 2010, http://money.cnn.com/2010/08/16/news/companies/ drug_recall_surge/index.htm.

Part II
Strategies, Frameworks, and Challenges for Health Analytics

5

Grasping the Brass Ring to Improve Healthcare Through Analytics: The Fundamentals

Dwight McNeill

The U. S. healthcare industry faces enormous challenges. Its outcomes are the worst of its peer wealthy countries, its efficiency is the worst of any industry, and its customer engagement ratings are the worst of any industry. Although the industry is profitable overall, ranking fourteenth among the top 35 industries,[1] it has had difficulty in converting these challenges into business opportunities to do good by improving the health of its customers while doing well for its stockholders.

One of the paradoxes of healthcare is that it uses science (a.k.a. analytics) more than any other industry in the discovery process—that is, to understand causes of diseases and develop new treatments. Yet there is a tremendous voltage drop in deploying this knowledge in the delivery of care and the production of health. McGlynn et al. reported in their classic paper "The Quality of Healthcare delivered in the United States" that patients received recommended care about 55% of the time and that these deficits "pose serious threats to the health of the American people."[2] This percentage is close to a coin toss. An example of this shortfall in translating research into practice is the mortality rate amenable to healthcare. Included in this mortality rate are deaths from diseases with a well-known clinical understanding of their prevention and treatment, including ischemic heart disease, diabetes, stroke, and bacterial infections. In other words the science is clear on what needs to be done, and the premature mortality rate from these diseases is an important indicator of the success on executing on the science. It turns out that the mortality rate of the United States is worst among 15 peer wealthy nations. In fact, it is 40% higher than the average mortality rate of the best five countries. This translates into 118,000 lives that would have been saved if one simply lived in these other countries.[3]

The reasons for the voltage drop are many. One is the predilection of highly trained and autonomous practitioners (doctors), who drive 80% of healthcare activity and costs, to rely on intuition rather than data to drive their decision making. This is related to not having information at their fingertips for decision making because it is either not there (research not digested) or is known but not findable (not digitized or available in electronic or paper records). For example, the human error rate in medical diagnosis is 17%.[4] It remains to be seen if the IBM *Jeopardy!* Watson application—incorporating natural language processing and predictive analytics for differential diagnosis—will reduce the rate, but it seems likely.[5] What is clear so far is that much of the voltage drop has to with the sociology, for example, changing cultures and behaviors, and not the technology. Going forward, analytics needs to make the case that it can produce compelling solutions to vital business challenges.

Delivering on the Promises

The promise of analytics in healthcare is huge. The McKinsey Global Institute states that "if U.S. healthcare were to use big data creatively and effectively to drive efficiency and quality, the sector could create more than $300 billion in value every year" through applications such as comparative effectiveness research, clinical decision support systems, advanced algorithms for fraud detection, public health surveillance and response, and more.[6] IBM estimates that if all the available IT and analytics solutions that it sells to health plans were fully and successfully deployed, a midsized health plan could net a potential $644 million annually in economic benefit.[7] And many other healthcare technology vendors offer the same proposition. For example, SAP, the largest business software company, states that "with the right information at the right time, anything is possible...and with real-time, predictive analysis comes a shift toward an increasingly proactive model for managing healthcare."[8]

It is now time to convert the promise of analytics and the potential benefits into practice and demonstrate results with quantifiable savings in terms of dollars and lives. Although analytics has been around a long time in various guises in healthcare, for example, informatics, actuarial science, operations, and decision support, a tipping point may have been achieved with the proliferation of big data, new technologies to harvest, manage, and make sense of it, and the acute need of business to achieve results and indeed transformation in the wake of the Great Recession and Obamacare. It may be a new ball game for analytics.

Fundamental Questions and Answers

If we boil it down to the fundamentals, businesses in all industries strive to accomplish two goals: Increase revenues and reduce costs. Figure 5.1 provides a healthcare value framework for these goals.

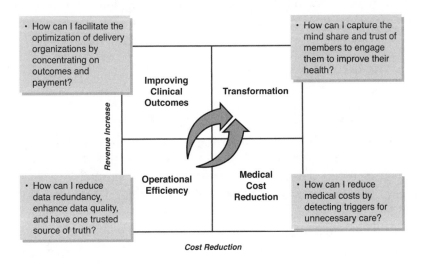

Figure 5.1 Healthcare value framework

On each axis, one for revenue increases and the other for cost reduction, more is attained by going up or to the right, respectively. In terms of cost reduction, gains can be made through operational efficiencies and through medical cost reduction. Clearly medical costs represent the majority of costs, and this is where the largest potential savings can accrue. Similarly on the revenue axis, gains can again be made through operational efficiency, but the larger area of opportunity is improving clinical outcomes. The cell with the most potential that combines optimization of both revenues and costs is transformation of the business to radically reposition the company for greater market opportunities. Examples of questions that address each cell of the matrix and require analytics support are provided in Figure 5.1. For the transformation cell, customer analytics using big personal data is the approach for knowing customers and providing them value. Another example is for operational efficiency, which addresses how to reduce data redundancy, enhance data quality, and have one trusted source of truth. The challenge for analytics and IT is to move out and upward from its usual focus on its own operational efficiency to providing value to the big challenges of the business.

So, this part of the book is devoted to strategies, frameworks, and challenges and includes six chapters that provide some fundamental answers to those wanting to step to the plate and get in the new game of health analytics. (Future chapters will provide how-to applications and best practices.) Some questions discussed in Part II are

- What is health analytics and what are the scope and various options?

Jason Burke, in Chapter 6, "A Taxonomy for Healthcare Analytics," asserts that the fundamental improvements needed in health and life sciences will only be realized via the deeper insights offered through analytics. He inventories the options in an analytics continuum that ranges from business analytics to clinical analytics. He includes the following five areas in his analytics taxonomy: clinical and health outcomes, research and development, commercialization, finance and fraud, and business operations. He catalogs an analytics "needscape," which is an inventory of analytics options that encompasses the various needs of healthcare organizations ranging from supply chain optimization to comparative effectiveness.

- What are the various ways to do analytics?

In Chapter 7, "Analytics Cheat Sheet," Mike Lampa, Sanjeev Kumar, Raghava Rao, et al., provide a "Rosetta Stone" for analytics to allow beginners to learn the language, including terms, acronyms, and technical jargon, quickly. They address the different types of analytics, for example, forecasting and text mining; analytic processes for enterprise scaling, for example, Six Sigma; sampling techniques, for example, random and cluster; data partitioning techniques, for example, test and validation set; a compendium of key statistical concepts, for example, correlation; modeling algorithms and techniques, for example, multiple regression; times series forecasting, for example, moving average; and model fit and comparison statistics, for example, chi square. All in all, it is a great reference for a quick lookup for those involved in analytics work.

- What is its value to the business and how is it determined?

Pat Saporito, in Chapter 8, "Business Value of Health Analytics," demonstrates that healthcare faces many challenges including relatively poor outcomes, inefficiency, and low customer satisfaction. Analytics offers a value proposition to use insights derived from data to solve some difficult business issues. But analytics is funded by the business and the Return on Investment (ROI) must be clear for business to invest in any endeavor. Saporito makes the case that analytics must be aligned with the business, first and foremost. She details a number of ways to prove the value of analytics, overcome biases about analytics, and change the culture to be more fact based in its decision making.

- How do I keep out of trouble with my lawyers about privacy concerns?

Thomas Davenport, in Chapter 9, "Security, Privacy, and Risk Analytics in Healthcare," makes the convincing case that the future of advanced analytics to meet evolving business needs that relies on deep and diverse data sets, often including personal data, can be put on a fast track or compromised, depending on approaches to identity protection and privacy. He suggests that companies that do it well may achieve an advantage in the marketplace. He presents the kinds of adjustments that leading payers, providers, and life science organizations are making to their information security and privacy practices. He also details how healthcare organizations are using risk analytics to bolster their security practices. He comments that analytic leaders are paying closer attention to data ownership and privacy and that lawyers rather than IT teams will determine how to interpret and protect privacy.

• What are some examples of analytics "secret sauce"?

Dwight McNeill in Chapter 10, "The Birds and the Bees of Analytics: The Benefits of Cross-Pollination Across Industries," addresses how healthcare can learn from other industries, including retail, banking, sports, and politics. McNeill asserts that industries have unique strengths and get proficient in associated "sweet spot" analytics. Other industries are blinded from them and their potential performance is constrained. He focuses on the following areas: Why analytics innovations matter; how to find and harvest analytics sweet spots; what best practices analytics should be adapted in healthcare; and how to put analytics ideas into action by understanding the innovation adoption decision-making process. He proposes seven adaptations that address seemingly intractable healthcare challenges, such as population health, patient engagement, and provider performance.

Notes

1. CNNMoney, "Top Industries: Most Profitable 2009," accessed February 28, 2013, http://money.cnn.com/magazines/fortune/global500/2009/performers/industries/profits/.

2. E. McGlynn et al., "The Quality of Healthcare Delivered to Adults in the United States," *New England Journal of Medicine* 348 (2003): 2635-45.

3. D. McNeill, *A Framework for Applying Analytics in Healthcare: What Can Be Learned from the Best Practices in Retail, Banking, Politics, and Sports* (Upper Saddle River, NJ: FT Press, 2013).

4. Agency for Healthcare Research and Quality, Diagnostic Errors, http://psnet.ahrq.gov/primer.aspx?primerID=12.

5. IBM, "Memorial Sloan-Kettering Cancer Center, IBM to Collaborate in Applying Watson Technology to Help Oncologists," March 22, 2012, www-03.ibm.com/press/us/en/pressrelease/37235.wss.

6. Basel Kayyali, et al., "The Big-Data Revolution in US Healthcare: Accelerating Value and Innovation," April 2013, www.mckinsey.com/insights/health_systems/The_big-data_revolution_in_US_health_care.

7. P. Okita, R. Hoyt, D. McNeill, et al., "The Value of Building Sustainable Health Systems: Capturing the Value of Health Plan Transformation," IBM Center for Applied Insights, 2012.

8. SAP, "Global Healthcare and Big Data," marketing brochure.

6

A Taxonomy for Healthcare Analytics

Jason Burke

As described in Chapter 1, "An Overview of Provider, Payer, and Life Sciences Analytics," by Thomas Davenport and Marcia Testa, the global healthcare ecosystem—healthcare providers, payers, and life sciences firms of all types—is undergoing a transformation. Though priorities vary across organizations and geographies between cost, safety, efficacy, timeliness, innovation, and productivity, one universal truth has emerged: The fundamental improvements needed in health and life sciences will only be realized via the deeper insights offered through analytics.

Most, if not all, of the analytical capabilities needed to drive systemic changes in healthcare have been available in commercial software for decades. Though a multiplicity of reasons exist why analytics have not been deployed more pervasively and comprehensively within healthcare, the reality is that most health-related institutions today have some limited analytical capability and capacity. As executives and leaders develop their respective institutional transformation plans, there is a need to consistently characterize and assess an organization's analytical capabilities.

Toward a Health Analytics Taxonomy

For organizations looking to grow their analytical competencies, one initial challenge is simply understanding the inventory of options. What are all of the ways that analytics might help transform the business, and how can priorities be developed against those options? What are the focus areas?

Despite areas of analytical progress in niche market topic areas,[1] a common taxonomy for health analytics has yet to emerge. However, some noticeable trends have surfaced:

- As illustrated in Figure 6.1, analytical applications in health and life sciences are increasingly being conceptualized as existing on a continuum between *business* analytics (e.g., cost, profitability, efficiency) and *clinical* analytics (e.g., safety, efficacy, targeted therapeutics).[2]

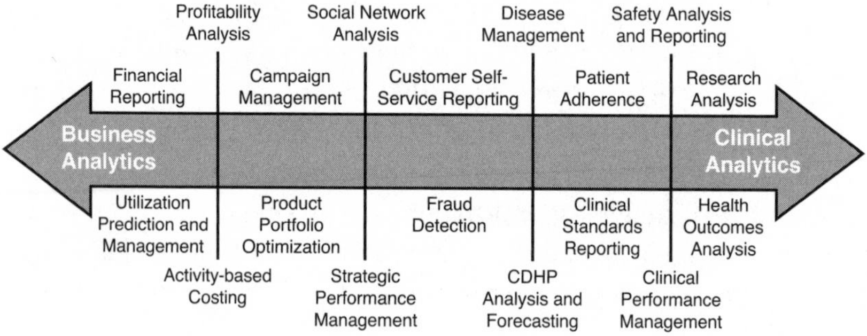

Profitability Analysis	Social Network Analysis		Disease Management	Safety Analysis and Reporting
Financial Reporting	Campaign Management	Customer Self-Service Reporting	Patient Adherence	Research Analysis

Business Analytics ⟶ **Clinical Analytics**

Utilization Prediction and Management	Product Portfolio Optimization	Fraud Detection	Clinical Standards Reporting	Health Outcomes Analysis
Activity-based Costing	Strategic Performance Management	CDHP Analysis and Forecasting	Clinical Performance Management	

Figure 6.1 Health analytics continuum

- Whereas organizations have created initiatives targeting the extreme ends of that continuum (e.g., an activity-based costing initiative at a hospital), the largest challenges still reside in moving toward the middle of the continuum: linking clinical and business analytics into a more comprehensive view of health outcomes and costs.

- To successfully link the business and clinical perspectives, data from all three traditionally "siloed" markets—care providers, health plans, and researchers/manufacturers—must be joined to produce a more complete picture of quality, efficacy, safety, and cost.

So in summary, the analytically derived insights needed to drive health industry transformations require industrywide collaboration around shared information and common analytical needs that link clinical and business concerns. A common analytical landscape, though not supported through industrywide consensus at the moment, can

a) form the basis of agreement within an organization for purposes of strategic planning and organizational development, and b) begin to offer a common understanding of the analytical underpinnings of meaningful health transformation.

Drafting a Health Analytics Taxonomy

At first glance, it may appear to be an impossible task: How can a single representation of analytical needs capture the breadth, depth, and diversity that currently exist within the global healthcare market. And in truth, it probably cannot. Yet despite differing market structures, business models, and incentives, most healthcare organizations have similar analytical needs: how to identify the best treatments, how to operate more profitably, how to engage customers more effectively, and so on. Though the motivations behind undertaking analytical initiatives may vary, both the analyses and their corresponding data are comparable.

At the highest level, we have observed five areas of analytical competencies that modern health organizations—including providers, payers, and life sciences organizations—are discovering will be needed to successfully compete:

1. **Clinical and health outcomes analytics**—These analytics are related to maximizing the use of existing treatments and therapies. For example, providers and health plans are both driven to ensure the best treatment is pursued for a particular patient, not just patients in general.

2. **Research and development analytics**—These analytics are related to discovering, researching, and developing novel treatments and therapies. For example, pharmaceutical researchers and providers need to know where potential clinical trial participants can be located to expedite new drug development.

3. **Commercialization analytics**—These analytics are related to maximizing sales, marketing, and customer relationship efforts. For example, both providers and health plans are motivated to

communicate more frequently and effectively with patients regarding products, services, and treatments.

4. **Finance and fraud analytics**—These analytics, which could be considered part of business operations (discussed next), relate to ensuring the financial health and stability of the organization. They are called out separately here due to the strategic role that claims, fraud, and risk play within healthcare markets.

5. **Business operations analytics**—These analytics are related to driving productivity, profitability, and compliance across the various business functions of an institution. For example, health plans and pharmaceutical manufacturers are motivated to ensure optimal operation of call center facilities, staff, and assets.

Across these five domains, a spectrum of analytical needs/scenarios can be described, as depicted in Figure 6.2. Each box represents a class of analytics-related needs and capabilities; taken together, an "analytical needscape" emerges that describes the overall landscape of analytical concerns that organizations need to address.

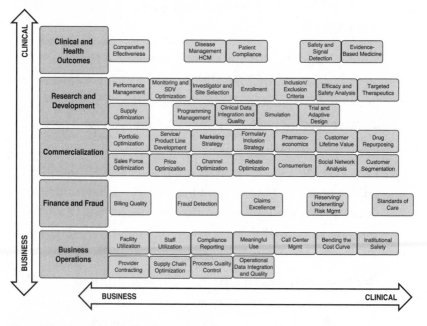

Figure 6.2 Health analytics "needscape"

Ultimately, organizations may differ in

- **What is included in the analytical needscape**—The list of analytical possibilities is large if not endless, and organizations may uncover opportunities for innovation in pursuing analytical scenarios not represented in Figure 6.2. The two most important aspects of building the needscape are a) accurately reflecting the needs of the organization, and b) reflecting those needs in a way that is generalizable to more than one market so that discussions around shared data and analytical models can proceed unimpaired.

- **Where certain analytical needs sit within the needscape**—Should "bending the cost curve" (a colloquialism regarding restructuring clinical and administrative cost models to reflect a post-reform U.S. business climate) be in the "Business Operations" or "Finance & Fraud" domain? In the end, it probably doesn't matter as long as the needs are represented.

- **The relative priorities within the needscape**—Organizations may competitively differentiate themselves based on business strategies that imply differing priorities across the various analytical scenarios. Alternatively, their business model may naturally reflect a bias toward certain analytics. These priorities may differ by functional business unit within an organization as well.

The construction of an analytical taxonomy provides a common communication vehicle by which organizational and functional leaders can map those aspects of analytical capabilities critical to the long-term health and success of their business.

Notes

1. For example, the detection and prevention of healthcare financial fraud, or the emergence of adaptive clinical research designs within clinical trials.

2. Jason Burke, "The World of Health Analytics," in *Health Informatics: Improving Efficiency and Productivity*, ed. Stephen Kudyba (CRC Press, 2010).

7

Analytics Cheat Sheet

**Mike Lampa, Dr. Sanjeev Kumar, Dr. Raghavendra,
Dr. Raghava Rao, R. Jaiprakash, Dr. Brandon Vaughn,
Vijeyta Karthik, Arjundas Kesavdas, Elizabeth Thomas**

Analytics offers any business a means of differentiating itself from and outperforming its competition, but for organizations just getting into the analytics arena there can be a steep learning curve. The recruitment of highly trained statisticians, econometricians, and data miners certainly enhances an organization's capability to build and deploy analytics. But often managers and executives find it difficult to communicate with analysts in a domain replete with acronyms, mathematical concepts, and buzzwords. The statistical concepts and distinctions behind advanced analytics can often be difficult to understand. What is the difference between statistics and data mining? What's a model? What's an ROC? What's AIC? Which is better, a high or low value? Making intuitive and relevant sense of it all becomes a big job. Where does one begin?

In this chapter, we take a step toward demystifying analytics by defining the basic and important concepts and processes. A wide expanse of knowledge is available on the Web in the form of tutorials, primers, and whitepapers on statistics, data mining, forecasting, and so on. So we provide brief descriptions and appropriate references so as not to reinvent the wheel. We present the material in a step-by-step manner to give abstract concepts firm grounding and as a series of tables to facilitate quick reference. We've tried to make this more than just an analytics glossary, but an analytics "cheat sheet" that enables beginners to get up to speed and provides a quick reference for the seasoned analyst.

Different Types of Analytics

Table 7.1 captures the most popular flavors of analytics in use within the business world today. While these types of analytics serve different purposes, there is considerable overlap in the use of techniques across the types; for example, correlation and covariance are common to all types of analytics, and regression-based techniques are popular in both statistics and data mining.

Table 7.1 Types of Analytics

Type	Description
Statistics	The science of collection, analysis, and presentation of data.[1]
Forecasting	The estimation of the value of a variable (or set of variables) at some future point in time as a function of past data.
Data mining	The automated discovery of patterns in data through the use of computational algorithmic and statistical techniques.
Text mining	The process of using statistical, computational algorithmic and natural language processing techniques to extract information from text in a manner similar to data mining.
Optimization	A discipline that uses mathematical techniques to answer questions such as, "what is the best that can happen?" Problems within optimization frequently involve finding optimal solutions while satisfying constraints.

Analytic Processes

Performing analytics involves various degrees of science and art. Individual analysts typically carry out analysis in a personal manner. However, scaling this out to a team can become difficult for managers. Consequently, adopting an analytical process framework is usually a good way to help standardize a team on analytics process while helping to spread analytics leading practice across the enterprise. Table 7.2 describes three common analytic processes that represent excellent, tried and tested methods. Note that when selecting an analytic process, it is important to evaluate the pros and cons of each methodology for your business's and teams' needs.

Table 7.2 Analytics Processes

Analytic Process	Description
SEMMA	Attributed to SAS Institute, SEMMA is an acronym that refers to the core process of conducting data mining. SEMMA stands for five steps:
	S = Sample: An optional step in which a representative subset of data is extracted from the larger data set. This step can also be used when resource constraints render processing against the larger data set unfeasible.
	E = Explore: A step in which the analyst can visually or statistically examine the data to uncover, for example, trends and outliers.
	M = Modify: A step in which data modifications such as binning and transformations can be applied.
	M = Model: A step in which data mining models are built.
	A = Assess: The final step in which the performance of models is assessed to test the validity of the model results.
CRISP-DM	Cross Industry Standard Process for Data Mining. A common and popular[2] data mining process model.
	CRISP-DM organizes the data mining process into six phases:
	• Business understanding
	• Data understanding
	• Data preparation
	• Modeling
	• Evaluation
	• Deployment
Six Sigma and DMAIC	Six Sigma[3] is a business management strategy that seeks to eliminate defects. Six Sigma employs statistical approaches to quality control to ensure that 99.99966% of subject outputs are statistically expected to be free of defects (no more than 3.4 defects per million produced).
	DMAIC[4] is an improvement methodology that when combined with Six Sigma provides an analytical, measurements-based framework to increase quality and reduce costs. DMAIC has five phases:
	D = Define
	M = Measure
	A = Analyze
	I = Improve
	C = Control

Data Scales

Understanding the different types of data scales is critical to ensuring you do not lose information or introduce erroneous signals into your data sets during data preparation and analysis. Table 7.3 provides a list of data types and what they mean.

Table 7.3 Scales of Data Measurement

Measurement Scale	Description
Categorical/ Nominal	Pure categorical measures of a variable; they have no size, order, or degree of differences between groups. They are typically labels or attributes (e.g., occupation, eye color, gender, fraud).
Ordinal	Ordinal variables are special categorical variables that enable ordering, but have no relative size or degree of difference (i.e., numerical operations such as addition/subtraction have no meaning). Examples would be "Like" survey items (Strongly Disagree, Neutral, Strongly Agree) and education level (elementary school, high school, some college, college graduate). It is important to note that for this type of variable, if numbers are used to represent the categories, then the numbers are just labels, and the difference between levels does not represent magnitude.
Interval	Interval variables are numerical attributes that have order, relative size, and degree of difference, but no absolute zero point. For example, most temperature scales are interval variables. It is possible to order values (e.g., 90°F is greater than 40°F). However, most temperature scales do not have an absolute zero (i.e., 0°F does not indicate the absence of a temperature; rather it is a distinct temperature value of significance). Interval measures are more common for manmade scales, while naturally occurring measures (e.g., weight) tend to be ratio scaled.
Ratio	Ratio variables are the same as interval variables except that ratio variables have an absolute zero point. The temperature scale, Kelvin, is a prime example of a ratio scale with absolute zero (equal to -273 degrees Celsius). Ratio scales enable comparisons such as 150 degrees is three times as high as 50 degrees. Most numerical variables (e.g., waiting time, salary (measured as a number), height, weight, etc.) would be on a ratio scale. In most statistical analysis, both the interval and ratio scales are treated as numerical variables, and some programs don't distinguish between them.

Sampling Techniques

Traditionally, statistics has employed sampling as a method of overcoming obstacles associated with not having an entire population available. Sampling helps to assemble a representative subset of individuals from a larger population to enable generalization. Table 7.4 lists common sampling techniques.

Table 7.4 Sampling Techniques

Technique	Description
Sampling	The process of selecting a subset of individuals from a larger population with a goal of using the sample data to generalize back to the general population.
Random sampling	The same as sampling, but with a probabilistic component to the selection process that allows every individual in the population a chance of being chosen.
Simple Random Sampling (SRS) with and without replacement	Same as sampling, but each individual has an equal probability of being selected. This is the most basic of all sampling designs and serves as the basis of more complicated random sampling designs.
	When SRS is performed with replacement, it means that the individual is placed back in the population and can be selected again. When SRS is performed without replacement it means the individual is not placed back into the population and cannot be selected again. This prevents multiple occurrences of the individual.
Undersampling and oversampling	Forms of sampling that adjust the number of individuals up or down in a sample. For instance, undersampling drops the level of a particular type of individual in a sample while oversampling increases the level of a particular type of individual.

Technique	Description
Stratified sampling	A technique that breaks a population into groups or subsets, known as *strata*. It is important that the groups are homogeneous (heterogeneous groupings would be better served by use of *cluster* sampling designs). For example, a population can be split by gender, with an SRS of males and a separate SRS of females. This sampling design can either select equal numbers of participants from each group (to ensure equal representation) or in proportion to the size of their strata (to mimic the population and ensure that dominant groups remain dominant in the sample).
Cluster sampling	A technique that takes intact groups of a population and randomly selects groups for study. It is important that the groups are heterogeneous. For example, a city can be divided into areas with no inherent characteristics.

Data Partitioning Techniques

During model construction, the data set will need to be divided into two or three different data partitions each with a different purpose as defined in Table 7.5. The partitions are constructed using sampling techniques as defined previously in Table 7.4. Partitioning should not be confused with sampling. The rationale behind partitioning is to build models that can be used for prediction. To do this one needs to train the model using as much historical data as possible; this data set is termed the *training set*. After training is complete, it is important to assess the model's capability to generalize to new data by testing the model on data it has not previously seen (i.e., was not included in the training set); this new data set is called the *test set*. This is the real test of what the model learned from the training set. Validation is the final step, in which a third data set is created to optimize and tweak the model should it be required. Note it is typical to see training and test sets only.

Table 7.5 Data Partitioning Techniques

Technique	Description
Data partitioning	The process of splitting the data set into two or three *partitions* for the purposes of building (training), testing, and validating statistical models. By randomly splitting data into these partitions, it allows the researcher to create "two or three samples" from one: one to build the model, one to test/improve the accuracy of the model, and a final one to provide the final validation of the improved model. Since the second and third data are not used in the model building, model accuracy shown when using the second data set establishes validity of the model. By randomly assigning the original sample into three partitions, a researcher has greater assurance that all three partitions are similar.
Training set	A partition of the full data set used to build/train models. A rule of thumb is to divide the sample into training and validation in the ratio 60:40, 70:30, or 80:20 (i.e., the training set is typically larger than the test set since it builds the model). If a third data set is needed, these rules of thumb are adjusted (either so that the test and validation set are the same size—both smaller than the training set—or where the validation set is slightly smaller than the test set).
Test set	A partition of the full data set (typically smaller than the training set) used to test how well the model can predict/classify new data previously unseen with a goal to possibly change and improve the model.
Validation set	A partition of the full data set used to test the final model. If the initial model is sufficient without the need for improvements, the test and validation sets are done as a singular action (i.e., one test set) instead of two independent processes.

Statistical Overview—Key Concepts

To perform effective analytics, a good understanding of key statistical concepts is required. Table 7.6 provides a description of key statistical concepts to kick start your learning. For example, understanding the difference between correlation and causation (a common mistake) can help you not make spurious claims from your results.

Table 7.6 Key Statistical Concepts

Key Concepts	Description
Causation	Relationship between an event (the cause) and a second event (the effect), where the second event is understood as a consequence of the first.[5]
Correlation	A statistical measurement of the relationship between two variables/attributes. Correlation is a necessary but insufficient condition for casual conclusions. The range is from -1 to +1 for traditional measures like r (Pearson Product Moment Correlation Coefficient).
	Correlation = +1 (Perfect positive correlation, meaning that both variables move in the same direction together)
	Correlation = 0 (No relationship between the variables)
	Correlation = -1 (Perfect negative correlation, meaning that as one variable goes up, the other trends downward)
Covariance	A type of *variability* that measures how much two random variables change together. Covariance typically is the foundation of other metrics (e.g., correlation).
	Uncorrelated variables have covariance = 0
	With correlated variables covariance is non-zero:
	Positive values indicate that the two variables show similar behavior.
	Negative values indicate that the two variables show opposite behavior.
Dependent variable or Response variable	The variable whose value you would like to predict or explain. For example, if a company wants to predict customer loyalty using various predictors, customer loyalty would be the dependent variable.
Independent variable or Explanatory variable	A variable used to help predict or explain a dependent variable. For example, if a company wants to predict customer loyalty, the various predictors (e.g., cycle time) would serve as independent variables. Analysis is required to assess the significance between independent variables and dependent variables as the independent variable may not be a significant predictor of the dependent variable. In some analyses (e.g., correlational/regression) if no correlation exists, best practice suggests the removal of the noncorrelated independent variable(s).

Key Concepts	Description
Null hypothesis	Under *Inferential* statistics, one wants to test a claim about a population. This claim is split into two competing hypotheses: the null and alternative hypotheses. The null hypothesis (H0) represents a "status quo" perspective that is often what we are attempting to disprove. The alternative hypothesis (Ha or H1) typically represents the change we want to show. For example, if a company wants to predict customer loyalty based on cycle time performance, a null hypothesis could be "Cycle time is not a significant predictor of customer loyalty" versus the alternative hypothesis "Cycle time is a significant predictor of customer loyalty." Data is collected and tested to see how "unusual" it is under the temporary assumption that H0 is true. Rare or unusual data (often represented by a *p-value* below a specified threshold) is an indication that H0 is false, which constitutes a *statistically significant* result and support of the alternative hypothesis.
p-value	When performing a hypothesis test, the *p*-value gives the probability of data occurrence under the assumption that H0 is true. Small *p*-values are an indication of rare or unusual data from H0, which in turn provides support that H0 is actually false (and thus support of the alternative hypothesis). In the example presented previously, if data is collected on cycle time and customer loyalty and a correlation test conducted, one might find a very small *p*-value, which would suggest that H0 is false (and thus that there is indeed a relationship between cycle time and customer loyalty).
Sample size	Is the number of observations from a population that have been selected and used for calculating estimates and/or making inferences.

Key Concepts	Description
Significance level or Alpha α	Is the amount of evidence required to accept that an event is unlikely to have arisen by chance (and thus contradicts H0). This value is known as the *significance level*, Alpha, α. The traditional level of significance is 5% (0.05); however, this value can be made more lenient or strict as needed. For example, stricter values can be used for situations that demand stronger evidence before the alternative hypothesis is accepted (e.g., α = 1% (0.01)). In the previous example, if actions from a hypothesis test involving cycle time and customer loyalty could cost a company millions of dollars if wrong, then the test might be conducted at a stricter value to be "more certain" of the results. A value of 5% signifies that we need data that occurs less than 5% of the time from H0 (if H0 were indeed true) for us to doubt H0 and reject it as being true. In practice, this is often assessed by calculating a p-value; p-values less than alpha are indication that H0 is rejected and the alternative supported. When H0 is rejected, the term *statistically significant* is often used as an abbreviated "catch phrase" of the final result.
Variability	Represents the consistency or spread of data referenced against a common point (typically the mean). This is often represented in terms of standard deviation, which acts like an "average distance from the mean." Other common measures include Range, Interquartile Range, and Variance.

Variable/Feature Selection

Statistical modeling is an iterative process that can involve hundreds or thousands of independent variables. Clearly, with so many variables that can impact the dependent variable, simple, parsimonious models help reduce complexity and present models with nonsignificant predictors removed. Table 7.7 describes techniques that help an analyst reduce the number of independent variables to a select critical few.

Table 7.7 Variable/Feature Selection Techniques

Variable/Feature Selection	Description
Forward selection – Stepwise selection	A regression-based model-building method to select variables that are predictive. It starts with an empty model and continues to add variables with the highest correlation to the dependent variable one at a time. Only those variables that are statistically significant are included. This is often used to build models consisting of only statistically significant predictors, and attempts to build the model without the aid of the researcher. In most research, forward methods are preferred over backward elimination, and in general automatic model selection should be used sparingly.
Backward elimination	Similar in concept to forward selection, but the opposite. This technique starts with all the variables included and systematically removes statistically insignificant variables from the multiple regression model. This is often used to build models consisting of only statistically significant predictors, and attempts to build the model without the aid of the researcher.
Principal component analysis	A technique used to reduce the number of variables that go into a model, principle component analysis (PCA) identifies those variables (known as the principal components) that account for the majority of the variance observed among the independent variables. This is often used as a variable reduction technique. Although predating factor analysis techniques, this technique is utilized less than the more robust factor analysis approach in modern practice.
Factor analysis	A statistical procedure used to uncover relationships among many variables. This allows numerous intercorrelated variables to be condensed into fewer dimensions, called factors. This is often used as a variable reduction technique. For example, if a researcher has more than 100 variables of interest to study, a factor analysis might enable a composite metric to be created that would capture the essence of the 100 variables in a handful of composite measures. The use of factor analysis is often preferred in modern practice over a principal component analysis.

Modeling Algorithms and Techniques

Table 7.8 describes some of the popular algorithms and techniques used in statistical and predictive analysis, though there are many others that are not listed.

Table 7.8 Modeling Algorithms and Techniques

Technique/ Algorithm	Description	Key Statistic(s)
Clustering/ Segmentation	A process that defines the allocation of observations, e.g., records in a database, into homogeneous groups known as *clusters*. Objects within clusters are similar in some manner, while objects across clusters are dissimilar to each other.	Model Specifics: Distance metrics for group structuring; R^2 is also available.
Discriminant analysis	Useful for (1) detecting variables that allow the greatest discrimination between different naturally occurring groups, and (2) classifying cases into different groups with "better than chance" accuracy (i.e., establishing group membership).	Overall Model: Chi-square test. Model Specifics: Correlations to define groups; Classification matrix that shows the accuracy hit rates for classification.
Linear regression	Uses one independent variable to predict a dependent variable. The type of relationship is considered linear. This model can be adjusted for polynomial patterns. If other nonlinear relationships are of interest, other regression techniques would be used.	Overall Model: ANOVA test (F-statistic) and R^2 value. Model Specifics: Regression coefficient (slope) and associated p-value.

Technique/ Algorithm	Description	Key Statistic(s)
Multiple linear regression	Uses multiple independent variables to predict a dependent variable. The type of relationship is considered linear. These models can be adjusted for polynomial patterns and interaction effects. If other nonlinear relationships are of interest, other regression techniques would be used.	Overall Model: ANOVA test (F-statistic) and R^2 value. R^2-adjusted sometimes useful. Model Specifics: Regression coefficients (slopes) and associated p-value
Logistic regression	Uses multiple independent variables to predict a binary, categorical dependent variable (e.g., yes/no). The type of relationship is considered nonlinear. The regression is based on likelihood of the binary outcome (presented in terms of log-odds, odds, or even probability of occurrence). Extensions of this technique can be used for categorical data with more than two values.	Overall Model: AIC, Chi-square (Hosmer & Lemeshow test). Model Specifics: Regression coefficients (log-odds, which can be translated back into odds or probability) and associated p-values (based on the Wald Chi-square test).
Nonlinear regression	Uses multiple independent variables to predict either a categorical or numerical dependent variable. This type of relationship is nonlinear and cannot be represented by simpler polynomial expressions. Thus, the shapes are other types of curved patterns (e.g., exponential, logarithmic, and so on). Traditional overall regression model statistics do not apply to such models.	Overall Model: SSE, MSE (a "pseudo" R^2 is possible). Model Specifics: Regression coefficients (meaning varies depending on the algebraic structure of the nonlinear model) and associated p-values.

Technique/ Algorithm	Description	Key Statistic(s)
Decision trees	Provide a treelike structure that models the segmentation of data into homogeneous subgroups. The decisions generate rules akin to if...then rules that classify/predict cases. Classification and Regression Trees (CART) are a type of decision tree technique that performs two-way splits on decisions. CHAID is a recursive decision tree technique (CHi-squared Automatic Interaction Detector). CHAID operates by performing Chi-square tests resulting in multiway splits on decisions.	Overall Model: Gini index; entropy measures. Model Specifics: Tree diagram indicating rules for splits and classification.
Analysis of Variance (ANOVA) 1-way	A statistical method to compare two population means using one factor.	Overall Model: ANOVA test (F-statistic). Model Specifics: Post-hoc tests or planned contrasts for group differences (along with associated p-values).
n-way ANOVA	A statistical method to compare two or more population means using two or more factors. Interaction (moderation) effects are possible to model in this.	Overall Model: ANOVA test (F-statistic). Model Specifics: Interaction Effects (if applicable), Main Effects, and Post-hoc tests or planned contrasts for group differences (along with associated p-values).
k-Nearest Neighbor	k-Nearest Neighbor (k-NN) is a classification algorithm that classifies single instances based on "similar" records surrounding the item of interest. The number of records used to classify the observation is represented by k.	Overall Model: Euclidean distances used to establish "neighbors" among the records in question. Model Specifics: Rules to assign a class to the record to be classified (based on the established distances and classes of its neighbors).

Technique/ Algorithm	Description	Key Statistic(s)
Association analysis/ market basket analysis	A method for discovering relationships between variables. Typically used to analyze transactional data in what is known as market basket analysis.	Measures of support and confidence are used to assess models. *Support* refers to how often a rule applies: So to use an example of, those who buy apples also buy pears; support represents the number of times apples and pears show up in transactions relative to the total number of transactions. *Confidence* refers to the reliability of the rule. So to use the previous example, confidence would measure the number of times apples occur with pears in transactions relative to transactions that contain apples.
Neural networks	Nonlinear computational models whose operation and structuring are inspired by the human brain. Neural networks model complex relationships between inputs and outputs to find patterns in data.	Overall Model: Classification Matrix and Error Report. Model Specifics: Weights that help detail the path model developed.
Time series	Regression-like model that uses time as a predictor while controlling for dependencies among the time-based measures.	MAPE, MAE, MSE (see Table 7.9).

Time Series Forecasting

Forecasting is an analysis technique that uses historical data to determine the direction of future trends. This type of analysis allows companies to have a long-term perspective on operations over time. Commonly used forecasting methods are presented in Table 7.9.

Table 7.9 Time Series Forecasting

Statistic	Description
Time series	Regression-like model that uses time as a predictor while controlling for dependencies among the time-based measures. An ordered sequence of data recorded over periods of time (weekly, monthly, quarterly, and so on—e.g., the daily closing values of Microsoft stock on the NASDAQ).
Moving average	A technique that averages a number of past observations to forecast the short term. When graphed, it helps in displaying the trends in data that are cyclical and in smoothing out short-term fluctuations. This forecasting technique is considered a basic approach to controlling seasonal fluctuation.
Weighted moving average	A simple moving average assigns the same weight to each observation in averaging, while a *weighted moving average* assigns different weights to each observation. For example, the most recent observation could receive a high weighting while older values receive decreasing weights. The sum of the weights must equal one.
Exponential smoothing	A method similar to moving averages except that more recent observations are given more weight. The weights decrease exponentially as the series goes farther back in time.
ARMA (autoregressive moving average)	A forecasting model of the persistence (autocorrelation) in a *stationary* time series. The model consists of two parts: an autoregressive (AR) part (which expresses a time series as a linear function of its past values) and a moving average (MA) part (which essentially is modeling/controlling for the noise/residuals in the model). ARMA models can be used to evaluate the possible importance of other variables to the system.

Statistic	Description
ARIMA (autoregressive integrated moving average)	An extension of the ARMA models that incorporate the element of nonstationary data. This technique helps in making any time series stationary by differencing; this step is especially important in forecasting because stationality is a necessary condition for all data modeling techniques. ARIMA techniques can be used to forecast nonstationary economic variables such as interest rates, GDP, capacity utilization forecasts for call center resource planning, and so on.
ARIMAX models	ARIMAX models are an extension of the traditional ARIMA model, which allows for the incorporation of several explanatory variables into the model design often as covariates. This helps in understanding and quantifying the effect each of the explanatory variables has on the dependent variable. Practically, inclusion of covariates can help in bringing some additional business aspects into the model (e.g., promotional data in a sales forecast).

Model Fit and Comparison Statistics

Once a model has been executed, how do you measure overall model performance? Table 7.10 describes the many statistics that enable an analyst to assess model fit and comparative performance; it also indicates whether you should look out for a high or low value. Again, this list is by no means exhaustive.

Table 7.10 Model Fit and Comparison Statistics

Statistic	Description	When To Use?	Best Value
AIC (Akaike's information criterion)	Measure of the amount of information lost when a model is used to predict new values.	Model Selection: Use to select between models.	↓
Classification matrix	Compares values in a test set against the values that a model predicts. The matrix shows how often the predictions were correct and how often they were wrong.	Assess the relative accuracy of models.	↓ (misclassifications)
Chi-square	A quantitative measure of the relationship between qualitative variables.	Helps in determining whether an association exists between two variables. Model Selection: A Chi-square value of zero indicates there is no relationship. Chi-square values are positive, and the higher the value the stronger the relationship, but the value does not indicate direction.	↑
F-statistic	The F-statistic used in an F-test (or Fisher's test) is a comparison of the spread of two sets of data to test whether the sets belong to the same population (i.e., if the precisions are similar or dissimilar).	Tests model fit by comparing the model against random chance.	↑
t-Test or student's t-test	Is a test statistic that tests whether the means of two groups are statistically different from one another.	Has a number of different uses, but the most popular use is for comparing the means of two groups.	↓

Statistic	Description	When To Use?	Best Value
GINI coefficient	Is a measure of inequality. Measures the un-evenness in the spread of values in the range of a variable.	Model Selection and Model Evaluation: Values range between 0 and 1 or 0 and 100 as a percentage. A value of 0 indicates perfect equality while higher values indicate greater inequality.	Problem dependent
KS statistic (Kolmogorov-Smirnoff statistic)	Gives a separation measure between models under consideration with the goal of finding an optimal model fit.	Model Selection: Use to select between models. Higher the value better the model.	↑
Lift	A measure of the expected benefit of using the predictive model compared to not using the model. (A *cumulative lift chart* is a plot of distribution of cumulative target records against cumulative distribution of total sample.)	Model Evaluation: The greater the area between the baseline and the lift curve, the better the model. Lift charts are good visual aids for measuring model performance.	Top left area of the lift chart.
MAE (mean absolute error)	Measures the closeness of a prediction to the actual outcome. Typically used in forecasting to compare the forecasted value to an actual value during model development.	Model Selection: Use to select between models.	↓
MAPE (mean absolute percentage error)	Measure of accuracy in a fitted time series value in trend analysis (should not be used if input data has negative or zero values).	Model Selection: Use to select between models.	↓

Statistic	Description	When To Use?	Best Value
MSE (mean squared error)	An "average" sum of squares, often found by dividing the sum of squares (SSE) by degrees of freedom. Provides perspective on amount by which the value implied by the estimator differs from the quantity to be estimated.	Model Selection: Use to select between models.	→
p-value	Probability of observing the data/test statistic (or something more extreme), assuming that the null hypothesis is true. Smaller values are an indication of *significant* results.	Use when testing a test statistic. When comparing its value against the desired level of alpha, α, the *p*-value must be less than α to be significant.	→
Regression coefficients	Represent a rate of change based on a regression model. Care must be taken in interpretation of coefficients in logistic and nonlinear models since regular regression model coefficients represent simple rate of change. Nonlinear model coefficients represent complex change that isn't linear (e.g., exponential rate of change versus simple linear).	Use in regression models when describing the relationship between an independent and dependent variable.	Typically values not close to 0.
RMSE (root mean square error)	Measures deviation from the actual values of the dependent variable and the predicted values based on the model. Often employed in regression models.	Model Application: Useful in constructing prediction intervals based on regression models.	→
ROC Index (receiver operating characteristic)	Curves that reveal how changing cutoffs affect the classifier's accuracy versus precision. The closer the ROC curve is to the upper-left corner, the higher the overall accuracy.	Model Evaluation: Values vary from 0.5 (discriminating power not better than chance) to 1.0 (perfect discriminating power).	N/A

Statistic	Description	When To Use?	Best Value
RSS (residual sum of squares)	Similar to SSE; a measure of the discrepancy between the data and an estimation model.	Model Selection	↓
R-squared (R^2)	Provides a measure of how well future outcomes are likely to be predicted by the model. A measure of the amount of variability accounted for.	Determining correlation. The higher the value, the better the model. (Note that this measure grows in size indefinitely simply by adding more variables. Use with caution.)	↑
R-squared (R^2) adjusted	Same as R-squared, but adjusted to account for the number of parameters. Scaled between 0 and 1; its value can go down as well as up.	Determining correlation.	↑
SSE (sum of square error)	The sum of squared distances between scores and their associated means. Often used to describe how well a regression model represents the data being modeled. Measures the amount of variation in the modeled values.	Model Evaluation	↓

Notes

1. American Statistical Association, http://www.amstat.org/careers/whatisstatistics.cfm.

2. Data Mining Methodology (August 2007), KDnuggets, Polls, http://www.kdnuggets.com/polls/2007/data_mining_methodology. htm; Data Mining Methodology (April 2004), KDnuggets, Polls, http://www.kdnuggets.com/polls/2004/data_mining_methodology. htm "What Main Method Are You Using for Data Mining (July 2002), KDnuggets, Polls, http://www.kdnuggets.com/polls/2002/methodology.htm.

3. "Six Sigma," Wikipedia, http://en.wikipedia.org/wiki/Six_Sigma.

4. "What Is Six Sigma?" Six Sigma.com, http://www.isixsigma.com/new-to-six-sigma/getting-started/what-six-sigma/.

5. "Causality," Wikipedia, http://en.wikipedia.org/wiki/Causality.

8

Business Value of Health Analytics

Patricia Saporito

The business value of healthcare analytics, and how to determine it, have been a top priority for those who espouse the importance of analytics. According to a study by Nucleus Research, there is an annual revenue increase possibility if the median Fortune 1000 business increased its usability of data by just 10%, translating to a $2 billion opportunity. The same study states that there is a 1000% return on investment (ROI) for organizations that make analytic investments. The spotlight on analytics continues to shine especially in light of increased expectations driven by the Big Data momentum and its least touted "V"—value.

Analytics in healthcare have been providing value but have largely been limited to siloed application areas, for example, clinical improvement, productivity management, and revenue cycle management. The value of analytics within each of these "silos" is often visible and mature, but their impact and contribution to the business are relatively immature from an enterprise perspective due to lack of an integrated enterprise approach. Therefore, integrating analytics across these domains and understanding the value leverage and correlations between them is still a rich area of opportunity. Organizations can realize this greater value by adopting a value management approach. However, culture and change management will be two significant hurdles to overcome as business users will naturally focus on their own departmental interests. These hurdles can be addressed by incorporating value management best practices and institutionalizing them in Business Intelligence Competency Centers.

Business Challenges

Healthcare has huge challenges especially in the United States. Healthcare costs continue to rise, and healthcare organizations continue developing strategies to address them. The Realization of Patient Protection and Affordability Care Act's (a.k.a. Obamacare) promise of containing costs is far off, and in the short term is creating turmoil in the market as providers, payers, and pharmaceutical companies review and revamp their business models to comply. Regardless of these changes, organizations must take a value-driven approach to analytics, one that states the value of analytics in the terms of the business or face the lack of funding. As healthcare organizations face competing requests for limited resources, investments in analytics are often overridden in favor of operational or transactional systems such as electronic medical records (EMRs), computerized physician order entry (CPOEs), and the like. Even when analytics are addressed, all too often they are operational analytics embedded within these operational systems such as a report of overdue accounts receivables. Analytics that span these transactional systems and can bring greater enterprise value get short shrift. Therefore it is critical that organizations have an analytic strategy that is business driven, is strategically aligned and shows value. This keeps the focus on choosing the right projects, clearly driving ownership and accountability for business results and delivery on commitments for these results. Communicating successful projects with proven value will ensure ongoing funding. Unfortunately, most organizations have not demonstrated compelling value propositions so far.

Value Life Cycle

Business owners need to make their needs known to their information technology (IT) partners and tie them to their business imperatives to get their share of funding and so that IT gets funding as well. This starts during the annual planning and budgeting process when the business submits requests for resources. Any significant project will require a business case to obtain resources and for prioritization. Following successful implementation of a project, the business and IT should also engage in a value realization analysis that demonstrates the actual value achieved from the analytic project. Organizations can use the value realization not only to communicate successful results but also for "cascading" funding to support future initiatives.

A value-based management approach to this value life cycle contains three phases: discovery—investing for impact; realization—delivering for business outcome; and optimization—governing for ongoing performance (see Figure 8.1):

- Value discovery develops a value-based business strategy and business case, enabled by technology and aligned with corporate objectives. It answers these questions: What are your *business imperatives* and how do you align your business and IT strategy? What is the expected impact of addressing these imperatives (*business case*)? What are the *right initiatives* to address the value creation opportunities, and how are they prioritized?

- Value realization develops transformational strategies to mobilize, deliver, and measure business results based on insights into leading practices or benchmarks. It answers these questions: How should the business prioritize, mobilize, and govern *programs to deliver value*? How can the business case be made *actionable* at the operating level, and how is value measured? How should the business govern, architect, deploy, and ensure quality of *master data*? How should the business govern, architect, and set up *an information framework* for business analytics?

- Value optimization assesses how the implementation and program compares to best or leading practices and recommends areas where the business can drive more value from current investments. Key questions addressed in this phase are: What is the *value realized* by your program and how can the business derive more value from existing investments? What insights do you have regarding the *total cost of ownership*? How does the *implementation* compare to best or leading practices? How can we utilize a periodic business *process "health check"* to continuously measure our progress versus our peers or analytic leaders? How can the business mobilize and govern a program to optimize success? What services make sense in a shared service center, and what is the right approach to setting up these *shared services*?

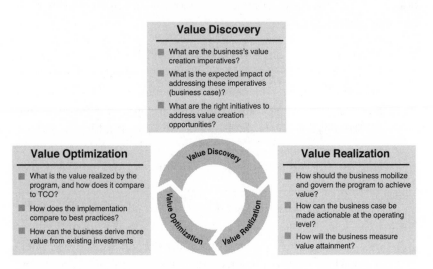

Figure 8.1 Value life cycle (Source: SAP)

More than 10,000 companies, both SAP customers and noncustomers, have participated in more than 30 business process benchmark surveys with almost 600 participating in Business Intelligence specific surveys. Based on this population, SAP has found that high adopters of this value management based approach achieve significantly more value than low adopters. These companies deliver twice as many projects on time and on budget, and show more than 1.5 times greater value than low adopters, regardless of industry (see Figure 8.2).

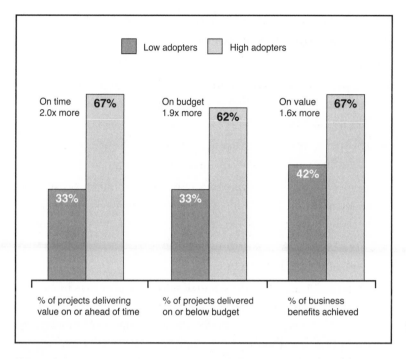

Figure 8.2 Low versus high value management adopters comparison

Healthcare Analytics Value Framework: Key Drivers

There are four key business drivers, or levers, to increase value in healthcare:

- Revenue growth:
  - Volume metrics measure new patient acquisition, as well as existing patient retention and growth.
  - Price metrics include increased prices per service as well as improved profitability per service and per patient.
- Operating margin:
  - General and administrative expenses metrics measure improving patient interaction efficiency and improving administrative service efficiency especially in HR and IT.

- Cost of care metrics include improving care service development and overall delivery efficiency
- Asset efficiency:
  - Property plant and equipment (PP&E) metrics focus on improving PP&E efficiency such as bed occupancy and/or room utilization.
  - Inventory metrics measure improving inventory efficiency including supplier management and leakage.
  - Receivables and payables management metrics improve receivables and payables efficiency, for example, percent of bad debt, percent of late charges, percent of A/R unbilled, and so on.
- Organizational effectiveness:
  - Organizational strengths metrics focus on improving management and governance effectiveness (e.g., business planning, business performance management) and improving execution capabilities (e.g., operational excellence, agility and flexibility, strategic assets).
  - External factors metrics include key macro- or microeconomic key performance indicators (KPIs) such as unemployment rate.

These levers impact three main business goals or areas: clinical performance, operational performance, and financial performance as shown in Figure 8.3. Each of these performance areas has several key business objectives:

- Clinical performance:
  - Improved clinical quality of care
  - Improved patient safety and reduced medical errors
  - Improved wellness and disease management
  - Improved patient acquisition, satisfaction, and retention
  - Improved provider network management

- Operational performance:
  - Reduced operational costs
  - Increased operational effectiveness and efficiency
  - Reduced inventory leakage
  - Improved provider pay for performance accountability
  - Increased operational speed and agility
- Financial performance:
  - Improved revenue
  - Improved ROI
  - Improved utilization
  - Optimized supply chain and HR costs
  - Improved risk management and regulatory compliance/ reduced fines
  - Reduced fraud or abuse leakage

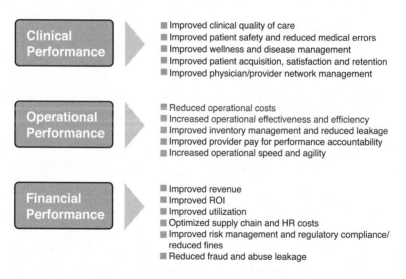

Figure 8.3 Healthcare performance management objectives (Source: SAP)

Validating Actionability and Measuring Performance Improvement with Key Performance Indicators (KPIs)

An effective way to develop a business case and also to validate analytic actionability and measurement is by analyzing and "diagramming" goals and objectives by asking the following questions:

- What business questions need to be answered?
- If those questions can be answered, what actions can be taken?
- Which KPIs need to be monitored to "move the needle" for improvement? What leading indicators should also be looked at for intervention before the resultant KPI? Are there any key correlation metrics you should also look at?
- What data are needed to support these objectives? How readily available are they? Where do those data reside; that is, what data sources do you need?

Figure 8.4 illustrates a framework used for working with a leading academic hospital system to define dashboards and reports to help manage nurse productivity and related analytic objectives.

Objective	Business Questions/Analysis	Actions	Measurable Results	KPIs	Data Sources
Improve HR life cycle	■ What is our turnover rate? (Internal vs. External, By staff type) ■ What is our employee satisfaction level? ■ How many open positions do we have? (By role) ■ What is our position vacancy rate? ■ How long does it take to fill these positions?	■ Determine reason for high turnover rate and take appropriate action ■ Determine reason for poor employee satisfaction and take appropriate action ■ Improve recruitment processes	■ Reduced turnover ■ Improved patient satisfaction level ■ Reduction in number open positions and time to fill ■ Percentage of RNs with bachelor's degrees meets guidelines	■ Turnover Rate ■ Patient Satisfaction Rating ■ Position Vacancy Rate ■ Position Time to Fill (Days)	■ Human Resource Information System ■ Patient Satisfaction Surveys

Figure 8.4 Objective actionability matrix framework applied to nursing productivity (Source: SAP)

Labor is one of the largest costs in healthcare. It is also one of the most critical resources in care delivery. Healthcare organizations face serious shortages of healthcare workers. Vacancy rate for nurses, imaging/radiology technicians, and pharmacists are over 10%; one in seven hospitals reports more than 20% of RN positions are vacant, with some reporting as high as 60%. Nursing productivity is a key analytic area that impacts all three performance areas—clinical, operational, and financial.

The organization already had a staff scheduling system in place that had some analytics, but it did not have an effective capability to integrate and analyze data across admissions, staff scheduling, human resources, and finance for use by the nursing unit managers. They originally identified about 20 KPIs they wanted to monitor; after our analysis we defined 30 metrics and expanded the scope of the data to be integrated.

As part of the business design process, we defined and validated five core objectives for Nursing Productivity:

- Enable compliance with staffing guidelines.
- Ensure patient safety.
- Improve the Human Resource life cycle (e.g., hiring, retention, promotion).
- Meet patient safety compliance guidelines.
- Attain revenue goals.

By using the dashboards and reports developed for this project the nursing unit managers were able to better anticipate and manage their staffing needs and costs. A "what-if" component allowed them to see the impact of one change on other areas—for example, the staff budget impact if they used more or less outside staff or increased permanent staff hours. As a result the healthcare organization elected to adopt a primary strategy of increased use of agency staff to reduce permanent staff overtime and burnout, resulting in reduced vacancy rates and related recruiting costs. They also saw an improvement in patient satisfaction as the permanent staff was not as stressed. All of these savings dropped to the bottom line. Without these expanded analytic capabilities to look across areas, they would have kept to stovepiped staff scheduling and staff budget analyses, a myopic approach, which was leading to increased permanent staff overtime and budget overruns, staff injuries, staff burnout and turnover, and decreased patient satisfaction.

Business Intelligence (BI) Performance Benchmarks for IT

In addition to revenue and expense impacts on the business, healthcare organizations also want to measure and improve their BI IT capabilities and overall performance to support the business. Sometimes IT attempts to justify analytic investments based solely on reducing Total Cost of Ownership (TCO); while TCO is important to reduce IT costs, it's only part of the value picture. The business case for investment must include business revenue and expense components, as well as TCO.

Once analytic investments have been approved and made, benchmarking is an effective way to measure performance both against internal best performance, and against external best in class in healthcare and across all industries.

Specific BI related objectives and related KPIs as part of an effective BI performance analytics program include

- Effectiveness:
    - Usage of BI by employees, executives, and external stakeholders (BI Adoption)
    - Level of insights generated through BI usage by business process (BI maturity)
- Efficiency:
    - BI project mix (e.g., reports, dashboards, semantic views)
    - Cycle times, reliability, and uptime
    - BI costs
- BI technology
    - BI technology leveraged for analytics, data warehousing, and data management
- Organization:
    - BI organization model
    - Level of centralization
    - Size of support organization

- Best practices adoption
  - Importance of best practices
  - Current coverage of best practices
  - Importance and coverage gap

Figure 8.5 shows a subset of metrics from the Americas SAP User Group (ASUG)-SAP BI Benchmark Survey, comparing a fictitious company against a benchmark peer group.

Performance Results–Metrics		Peer Group	
Metric	Company Value	Average	Top 25%
% Emp Building Reports In Off-line Tools	85.0	23.8	5.0
% Emp Building Reports on BI Tools	35.0	2.8	5.0
% Emp Analyzing Analytical Info on BI Tools	25.0	9.9	15.0
% Emp Executing Simple Queries on BI Tools	55.0	10.9	15.0
% Emp Receiving Analytical Info from BI Tools	45.0	28.6	50.0
Usage of BI to Manage Business Processes			
% Emp with Visibility into Strategic KPIs	25.0	10.0	15.0
Drilldown Levels	2	3	4
Implementation Time for a New KPI (in days)	3.0	21.5	2.0
% of Strategic KPIs Tracked Using BI S/W	22.8	31.5	50.0
SG&A (in %)	23.2	16.7	7.7
COGS (% of revenue)	72.1	68.1	56.8
Revenue per Employee (in 000s)	1111.7	436.0	600.0

Figure 8.5 Sample ASUG benchmark report (Source: ASUG, SAP)

Using BI Competency Centers to Institutionalize Value

BI Competency Centers (BICCs) can ensure that value management is part of every project and institutionalize it in an organization's overall BI program. Organizations have been forming BICCs (or Centers of Excellence—CoEs) to leverage best practices and improve operational effectiveness and efficiency to ensure business user satisfaction and demonstrate BI value. BICCs can report to IT or to business areas, often operations or administration; it is too early to see whether they are moving under chief analytic officers.

Regardless of where they report, BICCs play a key role in defining and executing an organization's BI strategy especially in demonstrating and communicating the value of analytics. These units have responsibility for several key areas of the strategy, including the development, documentation, and communication of the organization's overall BI strategy; development of the business requirements for a project and prioritizing all projects; defining the business case for each project and identifying supporting capabilities; defining an information taxonomy, architecture, and managing the technology tools; and finally managing the governance, program management, roadmap, measurement, training, and support (see Figure 8.6).

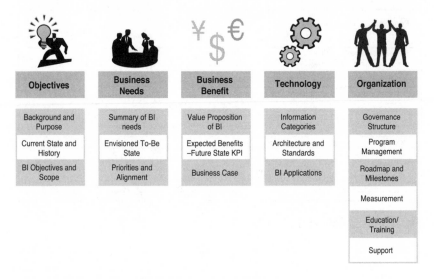

Objectives	Business Needs	Business Benefit	Technology	Organization
Background and Purpose	Summary of BI needs	Value Proposition of BI	Information Categories	Governance Structure
Current State and History	Envisioned To-Be State	Expected Benefits –Future State KPI	Architecture and Standards	Program Management
BI Objectives and Scope	Priorities and Alignment	Business Case	BI Applications	Roadmap and Milestones
				Measurement
				Education/ Training
				Support

Figure 8.6 BI strategy framework (Source: SAP)

Two key emerging roles that support value management are KPI analyst and value analyst. The KPI analyst helps define key performance indicators, leading indicators, and correlation metrics in the design phase. The value analyst helps define the anticipated ROI for the business case and validates attained ROI in the post-implementation value assessment. In smaller organizations these functions are often filled by the business analyst role, which can reside in the BICC or in business units. In larger organizations a value analyst may be part of a performance management function in finance or strategy areas. Both play a key role in the business benefit area of the BI strategy framework.

A third role that can contribute to value management is the communications analyst; this developing role can help broadcast analytic successes including quantitative value. Communications analysts are often part of the education and training function with the organization area of the BI strategy framework. They often use virtual communities of interest, or practice, to promote successes and increase employees' analytic engagement and adoption.

Business Performance Based Approach to Value

Using a value-based approach is a shift from a simple "on time and on budget" philosophy to a business outcomes and value-based one. Many companies focus on reducing IT and administrative costs, which are important but have far less impact on the business than revenue opportunities. Analytics-driven companies adopt a broader performance management approach and develop the following strategies and accountabilities:

- They demand that IT investments deliver competitive advantage.
- They focus change on high-impact results.
- They set ambitious business goals enabled by ambitious talent.
- They drive accountability through measurement.
- They embrace a culture of performance obsession.

We have found the following best practices are used to actualize value management:

- **Performance improvement**—They consistently and proactively measure and compare themselves internally and externally to identify new opportunities to gain more value from business processes.
- **Justification**—They have a formal process to justify investments that involves stakeholders across the business.
- **Value realization**—During project implementations, they design the solution to realize the value identified in the business case. They develop a key performance improvement

framework, complete with clearly identified ownership and accountability.

- **Business strategy-IT alignment**—Business process leads and IT professionals collaborate on everything from strategic planning to program execution to ensure that their objectives are well aligned.

- **Governance and portfolio management**—Executive leadership is engaged to help ensure project success.

- **Organizational excellence**—Programs are staffed and managed by a talented team that is measured based on the quantifiable business value its programs achieve.

Additional Resources

ASUG-SAP Benchmarking Forum is a facility to exchange metrics and best practices. It covers 26 business processes with 2,600 participants over 1,400 companies including heath care organizations. The BI survey has been conducted since 2007. Participation is free and open to both SAP and non-SAP customers. See http://www.asug.com/benchmarking.

Deloitte Enterprise Value Map is a one-page tool that shows the relationship between shareholder value and business operations. A healthcare-specific enterprise value map is also available. Search on "Enterprise Value Maps" at www.deloitte.com.

Global Environmental Management Initiative (GEMI) Metrics Navigator™ is a tool to help organizations develop and implement metrics that provide insight into complex issues, support business strategy, and contribute to business success. The tool presents a six-step process to select, implement, and evaluate a set of "critical few" metrics that focus on an organization's success. It includes worksheets, series of questions, or checklists for each step. (Note: While GEMI is focused on environmental sustainability, the metrics tools are excellent and apply across industries and processes.) Go to www.gemi.org/metricsnavigator/.

HIMSS Analytics and the International Institute for Analytics (IIA) have partnered to create the DELTA-Powered (Healthcare) Analytics Assessment, a new maturity model that assesses and scores the analytical capabilities of healthcare organizations. This survey and certification program provides a roadmap to gauge how organizations leverage data and analytics. Go to http://www.himssanalytics.org/emram/delta.aspx.

9

Security, Privacy, and Risk Analytics in Healthcare

Thomas H. Davenport

Information security, risk management, and privacy have always been hot topics across healthcare due to the high degree of government and industry regulation,[1] the sensitive medical information that is the industry's lifeblood, and the complexity introduced by significant outsourcing. While traditional information security software and hardware were designed to protect the huge data volumes and transactions required of deep analytics across payers, providers, and life science organizations, forward-thinking healthcare enterprises are more focused on redefining their approaches to identity protection and privacy to guarantee that when the use of analytics generates new opportunities, they can be put into action.

At the same time, highly public data breaches, stricter reporting requirements, and meaningful use reimbursement are forcing healthcare organizations to manage security and privacy in a dynamic and increasingly predictive manner. Analytics holds real promise in transforming how organizations both establish and report their security posture to key stakeholders, and could be the missing link to allowing organizations to market their risk posture as a competitive differentiator. Similarly, applying analytics should streamline compliance audits saving time and money. The applications with the most promise for IT and information security ("infosec") teams should focus on predictive analytics, correlation of physical and digital events, risk dashboarding, and assigning monetary value to data breaches.

This chapter first presents the kinds of adjustments leading payers, providers, and life sciences organizations are making to their information security and privacy practices to prepare for the expected explosion in the use of analytics across the enterprise. Second, the chapter details how healthcare organizations are using risk analytics to add a new dimension to their existing security practices, and in the process making their information risk posture more transparent to key stakeholders and regulators.

Securing Analytics

Information risk management at the atomic data level has been in place across payers, providers, and life sciences companies to varying degrees for years. Perimeter security (such as firewalls, intrusion detection systems (IDS), and intrusion prevention systems (IPS)), encryption, and data leak protection are common tools used to control the movement and use of sensitive information inside virtually every healthcare enterprise. Even the most basic security infrastructures are designed to scale to handle larger databases and transaction volumes, along with new transaction types, and adding analytical processes should not compromise the technological veracity of the common security infrastructure.

Instead, as the role of analytics is to generate new insights (information sets) from existing public and private data sets, leading organizations are reconsidering the current concepts of permission and privacy, differently within each sector. The age old practice of simply anonymizing input data sets will not satisfy the ever-changing customer privacy and government regulations:

> We've spent a lot of money on information security over the years, and feel pretty good on that front. It's really the legal and compliance side that has the hospital leadership most concerned. The idea of permission has to be completely rethought now if we're going to be able to use any of the results of the analysis we're now able to run. CISO, NY-based regional clinic.

Providers

Primary Data Types: Patient data (patient medical records, PII), clinical research data, financial and operational data

The historical motivators for information security and privacy protections inside provider organizations have been financial liability stemming from exposure of personally identifiable information (PII), and regulatory compliance. Among the three healthcare sectors, providers have the most difficult time with information security due to the different flavors of databases, desktops, mobile devices, diagnostic devices, and personnel (doctors, nurses). In addition, the high number of outsourced service partners the average provider organization shares data with inevitably leads to unintended information exposure.

Inadvertent data leaks have become the norm as permissioned access to data sets by insiders and outsourced service providers have become common. Cleanup costs of data leaks have grown considerably to include lost business to reputational damage, identity monitoring services for the affected patients, and forensic investigation costs to determine the exact source of the leak. While less common, malicious data breaches are certainly of more concern to most providers, and the costs can be considerably higher depending on the intent of the data thief.

While these issues may be exacerbated by heavier use of analytics, patient privacy will be the area where provider organizations will have to be sure that actionable insights can be used without the objection of patients. Traditionally, information security has been the domain of the IT staff. As new analytical techniques use more PII to generate more powerful insights, deciding what levels of permission must be secured to use those insights freely will need to involve the legal department more than IT. Providers who intend to become analytical competitors will need to shift their focus from securing their data to seeking permission to use the insights generated from those data.

Guidance: If analytical results are shared beyond normal boundaries (for example, partners, suppliers, etc.) and identity can be detected, the use of those results needs to be reviewed before taking action.

Trying to lock down all the tricky ways our medical staff use data and analytics to deliver better care is a losing battle for IT, and not their job. Instead, we initiated a dialogue with IT to be sure we don't get blindsided by a privacy suit. A great example is a recent effort to identify our most expensive patients over a long period of time, and then try to proactively work with those patients. Without proper permission from the patients to use their medical records for that purpose, combined with the financial information we have about them, we might run into a problem." Legal Counsel, Cambridge, Massachusetts-based hospital

Payers

Data: Patient data (patient medical records, PII, patient financial records), actuarial and rate information

Similar to provider organizations, payer organizations are built on collecting and making rate and coverage decisions using PII that allows them to predict insurance risks and set premiums accordingly. While analytics are mostly being applied at an aggregated data level, atomized data will be more and more accessible with analytics, making information security and privacy as critical as ever.

Similar to the provider community, leading payer organizations are involving legal teams more than IT teams as their ability to take action hinges on satisfying privacy rights among the insured community as well as to regulators. The bar for Health Insurance Portability and Accountability Act (HIPAA) compliance in particular will surely go up if consumers complain that new decisioning techniques can better assess an individual's risk, producing higher premiums for example. Claims of privacy violations could be a potent lever consumer groups could use to slow more efficient decisioning and classification.

Guidance: Negative reactions by consumers to rate increases and uncertainty around the ultimate outcome of the federal healthcare legislation are putting pressure on providers to price both existing and new policies correctly. Scenarios could be imagined where consumer health advocates could claim privacy over patient data used to drive toward rate increases or policy changes.

Life Sciences

Data: High-value research and development (R&D—e.g., drug discovery) data, clinical trial results

Unlike providers or payers, pharmaceutical and biotech organizations spend all of their information security resources on protecting highly proprietary research and development information, including clinical trial results, testing results, and drug formation results. Like providers, life sciences organizations are home to a variety of medical and research personalities, each of which has its own data recording and data storage cabinets. So, the focus tends to be on data leakage (securing the perimeter) and database security (access, encryption).

With analytics come more calls on large databases by numerous internal and external parties, which somewhat increases the security risk. But, even more than with providers, the security infrastructure inside most life science companies was built with scale as the number one priority. Since clinical trial data are largely aggregated and anonymized from the beginning, privacy is not an issue faced by life sciences organizations.

Guidance: Privacy is less of an issue here. Analytics will quicken the pace of proprietary information creation, and the information security infrastructure should grow to match that. Most often inside life sciences organizations, success is just having an adequate data inventory, and knowing where new proprietary information is being stored.

Risk Analytics

Most healthcare information security programs target the basics: internal awareness programs, desktop security, network perimeter protections, data leak prevention, authentication, and database security. Although mundane, leading providers are starting to package an aggregation of the basics into a security posture score and market this as a competitive differentiator. Patients like to know that their healthcare provider or insurer can reliably safeguard their information. And CEOs and board-level advisors like to know how much information risk the enterprise faces. Leading organizations are then calculating the return on investment in risk analytics as both internal (better security intelligence to predict threats and avoid costs) and external (marketing information safety results).

> We're right in the middle of trying to get budget to do some pretty cool things in terms of predicting when bad things will happen, mostly just on the inside from non-malicious users. We have the log files, and can react in near real-time, but we don't have the analytical tools to go to the next step. A by-product of this should also be a red-yellow-green type of indicator for the senior folks to know what the state of security is whenever they need it. CIO, Massachusetts-based multi-hospital group

In addition, while regulatory compliance takes constant attention (and always will), new acceptable use standards require security audits and tests to receive acceptable use reimbursement. Like PCI compliance among retail merchants, there now exists a real monetary incentive to continue to improve information security.

Predictive analytics,[2] correlation between physical and digital events, risk dashboarding, and valuing data breaches are the key areas where leading organizations are effectively applying analytics to information security today.

Predictive analytics is being applied to information security in isolated instances primarily to anticipate likely times when data breaches or data loss may occur. Generally, historical event patterns can help organizations project when internal processes are most likely to generate inadvertent exposure or loss. Another major area of focus is data exchanges with service provider partners. With enough historical data in hand, risk management teams can predict when a particular incident may be most likely to occur.

Correlating physical and digital events has been a focus of the security information management (SIM) vendors for years, but using advanced analytics to correlate more and more physical logs against digital events will continue. A big area of focus in many healthcare organizations is the intersection between the physical security and digital security groups. SIM software can now integrate and correlate, for example, real-time video surveillance records, network and database activities, and human resource records to make real-time incident response decisions.

Risk dashboards are maturing from the standard green-yellow-red approach to sophisticated pictures of the overall risk posture of an enterprise, complete with financial impact information that allows senior management to prioritize responses to significant information breaches and make proactive investments.

Calculating the monetary value of information security breaches will be a big domain of analytics. Information security has consistently lacked the capability to prioritize investment and responses because the value of the information assets being managed has been so difficult to estimate. With analytics, IT groups should be able to marry financial (value) data with breach information to be able to make more intelligent real-time decisions.

Compliance and Acceptable Use

HIPAA compliance garners the majority of the attention when it comes to regulatory compliance among healthcare providers, followed by Red Flag rules and the Health Information Technology for Economic and Clinical Health Act (HITECH).[3] Actual regulatory penalties suffered due to noncompliance are uncommon, but the perception of regulatory liability drives much of the spending in information security systems today. Will the application of analytics change this? Probably not. But, the leading organizations are certainly seeing an immediate payback to investment in risk analytics in the form of faster and more efficient reporting at lower cost.

Now, a renewed focus on data security has emerged due to its role in meaningful use reimbursement.[4] The rather vague requirements to conduct or review security and implement security updates to qualify for meaningful use reimbursement should generate a short-term bump in security spending to meet reimbursement requirements.[5]

Look to the Future

This chapter described how the growth of analytics is forcing all healthcare organizations to reconsider information security and privacy. Generally, existing information security infrastructures were built with scale in mind, and increased analytical activity should not affect an organization's security posture. Among providers and payers who are seeing the value of analytics, however, much closer attention is being paid to data ownership and privacy. Legal teams, rather than IT teams, will need to spend time understanding how to interpret and protect privacy when analytics create actionable opportunities.

While the biggest returns on analytics will never come in the area of information security, using analytics to improve an organization's risk reporting and overall understanding of risk posture will become a competitive differentiator in an era of increased regulatory requirements and incentives.

It is clear that the promise of analytics in healthcare could be handicapped without proper attention to security and privacy going forward. On the flip side, customers of all three healthcare groups expect intelligent security safeguarding their personal information that doesn't hinder their doctors, insurers, and pharmaceutical providers from caring for them quickly, efficiently, and cost-effectively. Analytics can help relieve this tension.

Notes

1. Examples are Health Insurance Portability and Accountability Act (HIPAA), industry-based Electronic Health Records (EHR) standards, Red Flags Rule, and Health Information Technology for Economic and Clinical Health Act (HITECH).

2. For a complete review of data types and correlation techniques, see "Security Analytics Project: Alternatives in Analysis," by Mark Ryan, Secure-DNA.

3. The Health Information Technology for Economic and Clinical Health Act (HITECH) is a new federal privacy and security mandate regarding patient information, including mandatory notification of individuals whose information is breached, that was included as part

of the American Recovery and Reinvestment Act of 2009 (ARRA), signed into law by President Obama February 17, 2009. A major change is that the new legislation generally requires covered entities and business associates to disclose to their patients any breach involving a patient's protected health information (PHI).

4. See *E-health Records: Getting Started with Meaningful Use* by Anthony Guerra for a summary of acceptable use reimbursement requirements related to data security systems.

5. See *2010 HIMSS Security Survey*, page 3, for a review and evaluation of risk assessment requirements in connection of with meaningful use reimbursement.

10

The Birds and the Bees of Analytics: The Benefits of Cross-Pollination Across Industries

Dwight McNeill

Healthcare can learn a lot from other industries. Industries develop deep and unique strengths in certain areas and get proficient in the associated analytics. Other industries that do not experience the same set of forces do not develop these analytics capabilities. They are blinded from them and their potential performance is constrained. This chapter provides a guided tour of industries that are somewhat mysterious to healthcare, including retail, banking, politics, and sports.[1] The express purpose is to harvest some ideas and build a bridge to adapt them in healthcare. Indeed, the best analytics from these industries provide insights to address some of the most intractable challenges in healthcare.

This chapter is about discovering ideas in faraway places and building bridges to welcome them home. It addresses four areas:

1. Why analytics innovations matter
2. The discovery process to harvest analytics sweet spots from the four industries
3. The distillation and interpretation of the sweet spots into the most compelling healthcare analytics adaptations
4. A roadmap for putting ideas into action and a model to evaluate the adoptability of the healthcare adaptations

Why Analytics Innovations Matter

Analytics in healthcare is a paradox. On the one hand, healthcare is immersed in analytics. It is far ahead of other industries in using science, for example, to understand diseases and develop new cures and treatments. But there is a significant voltage drop between the science and its application in practice. U. S. healthcare faces major challenges, not the least of which are its low standing in the world on key health outcomes, efficiency, cost, disparities, and affordability. And research shows that the odds of getting the right medical service are just a bit higher than 50/50.[2]

The fact is that we know how to address these challenges, and analytics can be a tremendous support; but there's a blockage. This is due to a number of factors, including the seemingly overarching prerequisite to digitize the business before anything else can be harvested from analytics. This forestalls other forms of analytics that can lead to real benefits for business today. But it is more complicated than that. There is an amalgam of contributing barriers, including the lack of good coordination among actors in the ecosystem, a perverse payment system that does not reward value, complex products and services, ambiguity about who the customer is, professional autonomy, convoluted market dynamics, multiple vested and powerful interests, and a pervasive, risk-averse culture.

In spite of these challenges, the field must innovate to make change happen. Innovation is critical to the success of any business. A recent MIT survey of 3,000 executives across many industries and countries found that the top business challenge is "innovating to achieve competitive differentiation."[3] This was far ahead of the usual business challenges to grow revenue and reduce costs. Innovation rises to the top at this time because the Great Recession required businesses to address all possibilities for cost reduction, and many have developed lean organizations. Businesses recognize that they need to grow the top line, and not just by extending the old line. And to grow the top line, they have to transform the business.

This transformation imperative is especially true in healthcare. For example, health insurers face commodity prices for premiums, the demand for transparency, and a flip of the business model from business-to-business to business-to-consumer, among other pressures related to healthcare reform. Similarly, providers are facing market and government pressures to improve outcomes, lower waste, and change the underlying revenue model from fee-for-service to global payments. In response, the leading companies are dramatically changing their identities. For example, some health insurers are becoming health companies, and others are viewing claims processing as one of many product lines as they become information companies. The industry archetypes are eroding, and innovation is charting a new path.

Innovation starts with fresh ideas. Ideas matter. They are the seeds from which innovations grow. Innovating to achieve competitive differentiation is the top business challenge today. And analytics is the high-octane fuel to power innovations to achieve business breakthroughs.

Ideas can come from comfortable places and unfamiliar places. This chapter concentrates on the latter, not because they are better, but because they are often ignored. They are ignored because they do not necessarily fit our beliefs about how the world works. We tend to seek out information that confirms our positions and ignore the rest, what is referred to as confirmation bias. So outside-in thinking has an inherent hurdle at the outset. But ignoring outside-in thinking can lead to blind spots. For example, legions of dedicated professionals concentrate on the problems within healthcare with the intent to improve them. The blind spot of this process might be characterized this way: "Removing the faults in a stage-coach may produce a perfect stage-coach, but it is unlikely to produce the first motor car."[4]

Discovery of Healthcare Adaptations

The process for discovering ideas from other industries is depicted in Figure 10.1. Note that this approach is based on the firmly held belief that analytics succeeds when it responds to the needs of the business and not when analytics *answers*—that is, technologies and methodologies—are in search of business *questions*.

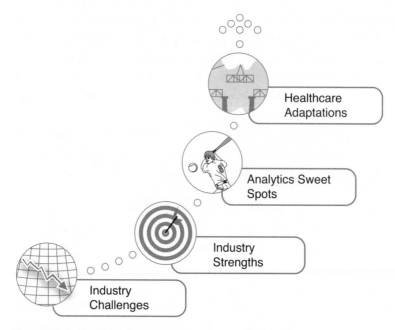

Figure 10.1 Healthcare adaptations discovery process

Industry Challenges

The first stage in the discovery process is industry challenges. This stage seeks to understand the industry, goals, context, challenges, and drivers. These ingredients mix together in a crucible that determines how an industry needs to perform to succeed by developing and honing areas of strength. For example, the banking industry suffered a disastrous plunge in trust and revenues due to its irresponsible subprime mortgage lending practices. This not only contributed to the Great Recession, which cut demand for its products, but also led to onerous regulations and oversight that constrained the business, especially in the area of lending. It subsequently developed greater strengths in risk assessment.

The industry challenges are unique to each industry and are summarized later in this chapter. However, there are challenges common to all the industries that are particularly acute at this time and go beyond generic business challenges, such as increasing revenues, reducing costs, improving the balance sheet, paying attention to customers, and so forth.

- **Great Recession**—This was not a typical, cyclical recession. In fact, it was the worst economic period since the Great Depression. It shattered customer purchasing power and the demand for products and services.

- **Hypercompetition**—This usual feature of business was put into high gear due to the Great Recession and the need to capture dwindling purchasing dollars. It makes demands on analytics to gather and analyze more diverse and larger volumes of data to know the customer more fully. This is most evident in the retail industry, where personal data can help predict what a customer wants to buy before the purchase.

- **Customer power**—Customers are taking the upper hand in their relationships with business because they have the data at their fingertips, in their smartphones and tablets, and they use them to make more informed purchasing decisions. The whole process of buying has been accelerated, yet made more deliberate, because of the availability of price information from multiple sellers on the Internet. Similarly, customers want to do business with companies "their way," on their preferred devices, with access to services 24/7, and expect the business to provide flawless service across channels. Finally, customers want information that makes sense to them and on their terms, which is often obtained from peers and not from marketing departments or government agencies.

- **Transformation**—The business race continues to have no finish line. But the difference is that the very nature of the business has to evolve and the pace of change is accelerating. The tried-and true ways of succeeding in banking, retail, healthcare, and even sports are up for grabs. The need for transformation creates the appetite for innovation.

- **Clicks**—Clicks are the sounds of doing business on computing devices. Clicks are challenging the bricks. What could be more indicative of shifting paradigms than the collapse of the structures in which people do business (bricks)? Mobile "rules" for now because it is seemingly a new organ of the body that offers new functionality, is integrated perfectly, and is appreciated. It is almost like a seventh sense for humans.

Industry Strengths

The second stage of the discovery process is the understanding of industry strengths. All industries have unique strengths that, if exploited, can drive business breakthroughs to beat the competition. The unique industry strengths, distilled to their essence, are summarized here:

- For retail, it is clear that the overarching strength of the industry is to acquire, retain, and optimize customers. Marketing's customer analytics are woven into this strength.

- For banking, the unique industry strength is understanding and minimizing risk. As mentioned earlier, the industry needed to clean up its lending practices for its very survival. Risk assessment has always been a core element of the industry, but it needed to get a lot better to reduce loan defaults without crimping the volume of loans. This became an industry imperative that necessitated the refinement of the industry strength.

- For political campaigns, it is the laser focus on finding, energizing, and persuading voters. This strength was made more powerful through data-driven innovations.

- For sports, it is its ability to engage fans through the full transparency of detailed performance data on its athletes and teams. Athletes are acquired, fired, and improved based on the data. The paradox for healthcare is that sports fans know much more about the athletes who entertain them than patients do about the doctors who make life-and-death decisions about them.

Analytics Sweet Spots

The third stage of the discovery process is identifying the analytics sweet spots that correspond to the industry strengths. The concept of a sweet spot comes from sports. It is the place where a combination of factors results in the maximum impact achieved relative to a given amount of effort. In baseball, for example, it is that place on the bat that produces the most powerful hit. The analogue in analytics is a solution that provides the most power to make the industry strength as successful as possible. In the case of banking, the analytics sweet spot, matched to the industry strength, is a refined FICO score that assesses the capability of a borrower to fulfill promises to repay a loan.

Each industry pushes the envelope in its use of specific analytics, not necessarily because it has more technical sophistication, but because the specific goals, purposes, pressures, and culture of that industry are unique and require better flowering of certain analytics tools. These are the sweet spots that we want to translate and adapt for healthcare.

There are cross-industry analytics themes that shape the manifestation of advanced analytics generally and have a strong influence on the analytics sweet spots for each industry.

Modeling Behavior Change

Predictive modeling to change behavior is a powerful, advanced analytics method used across the industries. Most of the breakthrough applications of predictive modeling across industries focus on understanding and changing behaviors of customers. Examples include the following:

- **Retail**—Determine the probability that a woman is pregnant and her estimated delivery date.
- **Banking**—Determine the likelihood of divorce as a major predictor of loan default.
- **Political campaigns**—Determine the messages most likely to convert an undecided voter.
- **Sports**—Determine what attributes of players contribute to team wins.

It's the Customer, Stupid

It goes without saying that customers are what make businesses thrive or die. Businesses can be distracted and focus on other priorities but do so at their peril. The pathway to growth is realized by understanding customers and responding in ways to earn their business. The analytics combo of predictive modeling with "boundless personal data" allows unprecedented views into what makes customers tick.

Boundless Personal Data

We live in a surveillance society. There is a huge business and government appetite to know everything about us. For example:

- **Retail**—"If you use a credit card or a coupon, or fill out a survey, or mail in a refund, or call the customer help line, or open an e-mail we've sent you or visit our Web site, we'll record it and link it to your Guest ID...and we can buy data about you... (such as) charitable giving and cars you own.[5]

- **Politics**—"The campaigns spent over $13 million on acquisition of data like whether voters may have visited pornography Web sites, have homes in foreclosure, are more prone to drink Michelob Ultra than Corona or have gay friends or enjoy expensive vacations."[6]

- **Military**—The National Counterterrorism Center can use any information the government has collected on you, including "flight records, lists of Americans hosting foreign-exchange students, financial records of people seeking federally backed mortgages, health records of patients at veterans' hospitals... this obscure agency has permission to study [any database] for patterns."[7]

Most of the industries reviewed are using boundless personal data to feed their customer analytics engines. The new data usually come from outside the industry including "open" public databases, data "snatchers," and Web click trackers. The large increase in the diversity and volume of personal data, in combination with other analytics methods such as predictive modeling and technology game changers, has been a significant factor in solving business problems and demonstrating the value of analytics across industries.

Big Data Promises

The promise of big data is great and alluring. McKinsey & Company proclaims that "it will become the basis of competition, underpinning new waves of productivity growth, innovation, and consumer surplus."[8] Boundless personal data is a piece of it. But big data goes beyond that and looks to extract meaning from every digital signal that is emitted. It is likely that harnessing big data will lead to a new world neural system that can measure almost anything. For the moment, it appears that the technology "hows" are ahead of the business "whys" and "whats." It is unknown how this revolution, like the Internet revolution before it, will play out and when the big promises will be fulfilled.

Technology Game Changers

Technology advances facilitate the use of data for analytics. Three game changers have been influential:

- **Clicks**—The Internet has transformed the way businesses communicate, market, do commerce with customers, and collect data about them. One example is the ability to do randomized trials, or A/B testing, of alternative Web site features—for example, how to get the most contributions during a political campaign—on large samples and virtually instantaneously.

- **Mobile**—Earlier we described the challenges and opportunities of mobile. It is seen as the preferred platform for customer communications across industries.

- **Hadoop**—Hadoop is an open source software framework that allows for the distributed processing of large data sets across clusters of computers using a simple programming model. It is less expensive and easier to get up and running than commercial applications; for example, it is cloud based and works across hardware. It has greatly facilitated the analysis of boundless data.

Looming Privacy Concerns

As boundless personal data increases the utility of analytics to address business needs, it also runs the risk that the "creepy factor" will stop it dead in its tracks. Most of these data are collected and used without the consent of the individual. For example, personal data are collected from children as they traverse the Internet and then used for tailoring marketing messages to them. This concern about privacy is acknowledged but largely ignored, and the response is often to deny its existence. The data are valuable and, while the gate is open, there are few restrictions on their use. But a few breaches of privacy might bring on a spate of consumer complaints and Congressional action.

Sociology, Not Technology

All the analytics sweet spots across the industries are successful because they complement and support important business needs and strengths. Excellent analytics methods can be developed in a bubble outside the realities of the business. But unless they are used to solve business problems, they collect dust and are an expense and not an investment. Making things happen/change/succeed is what business is about. It's not about the technology; it's the sociology of getting things done. What is clear from the analytics sweet spots is that the bull's-eye for the value proposition for analytics is understanding business challenges and strengths and providing tools and expertise in the right way and at the right time to support the business.

Analytics Adaptations for Health and Healthcare

The final stage in the discovery process is the translation of industry analytics sweet spots into potential adaptations in healthcare. It is hard to imagine, on the face of it, how banking is anything like healthcare. The task involves connecting dots between the industries to arrive at some revelation. It involves some logic, but is mostly about creativity. Creativity often leads to an epiphany.

For example, in the banking case, and reflecting on the industry strength in risk assessment, one ponders what it is about assessing customer capabilities and risks that might apply in healthcare. What emerged is that one of the most ingrained and intractable issues in healthcare is getting people to be active co-producers of their own health and thereby improve outcomes. People's behavior is the biggest influencer of health functioning followed by many social determinants. Can the risks, capabilities, and barriers to the fulfillment of doctors' orders, prescriptions, and plans that rest with the individual be measured and then managed better? This healthcare adaptation, called the Health Improvement Capability Score (HICS), is one of the seven healthcare adaptations that are distilled from the discovery process.

The seven adaptations, listed in Figure 10.2, cluster into three important areas of health and healthcare, including population health, patient engagement, and provider performance. The adaptations for population health include obesity detection, well-being, and my dashboard. For patient engagement, the adaptations include radical personalization and capabilities and support. Provider performance includes the adaptations of peer-to-peer and team centered.

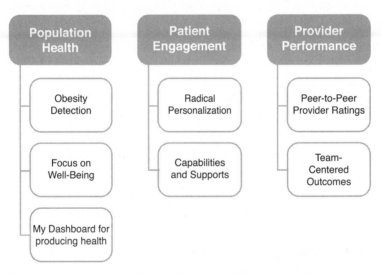

Figure 10.2 Seven health and healthcare analytics adaptations

Another example of an adaptation for healthcare comes from retail, banking, and politics...*to reduce obesity*. The goal is to reduce the incidence of diabetes through early identification of individuals with premorbid obesity, which is followed by targeted intervention programs to reverse weight gain. Many of these individuals will not have received sickness care, or might have received it for other reasons that did not include a risk assessment for diabetes. Therefore, there might not be any relevant data included in a traditional medical record or claims form that speaks to obesity. Retail, banking, and politics have multiple data sources and sophisticated analytics to identify likely buyers, uncommitted voters, or probable defaulters, and they have programs to sell, convince, or reject those so identified.

The analytics adaptation is straightforward. The determination of the degree of obesity is easily determined by a simple calculation of two variables: height and weight. The data are available from firms that aggregate publicly available data. The analytics methods include predictive modeling and the collection and integration of external data. The major challenges are belief systems that question the utility and appropriateness of external data and whether interventions intended to reverse obesity are effective.

Another example comes from sports—to improve provider outcomes. In sports, there is a long tradition of measuring the performance of athletes and making them transparent. One could focus on the paradox that consumers know much more about their sports heroes than about their doctors, though they should know more about a doctor's "batting average." But we take a different view. Performance measurement of players has been more about entertainment than about winning games. The pressures on the business of sports have moved the measurement to outcomes (winning) and how the combination of player attributes and individual performance contributes to team wins. Indeed, the sports industry now realizes that after decades of concentrating on individual performance, the business needs to make teams work better.

The analytics adaptation is to develop the ability to measure clinical performance at the care team level and demonstrate its superiority in driving performance improvements and outcomes relative to existing organizational levels of aggregation, such as a hospital or medical group. The measurement should dovetail with the emerging reality that care, especially for the chronically ill, requires a well-functioning team of physicians, other caregivers, communities, and patients. The measurement might well influence the management of care to make the components work together for the good of the whole.

The analytic methods include predictive modeling, external data collection, data integration, and dashboards. There are major challenges, including persistent resistance to performance measurement at a granular level and the belief that individual physicians are the key ingredient in producing patient outcomes.

Putting Ideas into Action

It is important to generate extra-industry ideas and to show the way across the bridge to healthcare. Ideas are fragile. Eggars and O'Leary think of ideas as seeds. Harkening back to a biblical parable, they note: "Some seeds land on rocky soil. Others are eaten by birds, and some sprout only to be choked by thorns. Only through a fortuitous combination of sun, soil, and water will a seed grow into a plant and bear fruit."[9] And progressing ideas to the endpoint of making a real difference is a long journey, and not all ideas deserve the passage.

The endpoint is embedding analytics into the ongoing operations of an organization. The pathway includes six stages. We have discussed the idea phase. The second is the design stage, which involves a plan of action and the justification for implementing it. The third stage is making the decision to adopt the plan, which is discussed in the next paragraph. The next stage is execution. Goethe noted in the eighteenth century, "To put your ideas into action is the most difficult thing in the world." The next stage is results where the performance of the innovation is evaluated. Unfortunately, most innovations fail from innumerable snags in the delivery of the program. Finally, the last stage is reinvention. A successful innovation changes and adapts through a learning and improvement process.

The adoption decision is complicated. Fortunately, there is body of knowledge that is represented in The Innovation Adoption Factors model (see Figure 10.3), which describes six areas or domains that need to be considered and mastered to persuade individuals to make the decision to adopt an innovation. The model is composed of two halves. On the left side is the idea stage and on the right is the design stage. The idea stage includes the domains of innovation receptivity, ideation maturation, and urgency/timing. The design stage includes the domains of the innovation's attributes, the organization's capabilities, and the readiness to position the innovation for adoption.

In the book, the six domains are decomposed into 18 factors. The factors become the basis for a guidebook that can be used for evaluation and planning for decisions about innovation adoption. A case study illustrates the scoring, management response, and improvements to the innovation process to make the innovation acceptable.

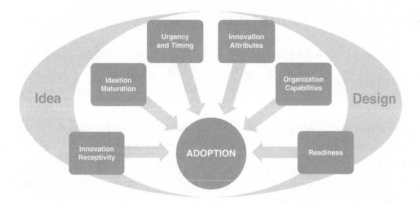

Figure 10.3 Innovation Adopters Factors model (Adapted from Rogers on diffusion theory,[10] Eggers and O'Leary on getting things done in government,[11] Kingdon on agenda setting,[12] Hasenfeld and Brock on policy implementation,[13] Pressman and Wildavsky on program implementation,[14] Gladwell on the tipping point model of change,[15] and Plsek[16] and Stacey[17] on complexity theory.)

Notes

1. The chapter is a brief summary of the book *A Framework for Applying Analytics in Healthcare: What Can Be Learned from the Best Practices in Retail, Banking, Politics, and Sports*, by Dwight McNeill and published by FT Press.

2. E. McGlynn, S. Asch, J. Adams, et al., "The Quality of Healthcare Delivered to Adults in the United States," *New England Journal of Medicine* 348 (2003): 2635-2645.

3. "Analytics: The New Path to Value," a joint MIT Sloan Management Review and IBM Institute for Business Value study, Massachusetts Institute of Technology, 2010.

4. Edward de Bono, BrainyQuote.com, Xplore Inc., 2013, www. brainyquote.com/quotes/quotes/e/edwarddebo389925.html (accessed 3/29/13).

5. Charles Duhigg, "How Companies Learn Your Secrets," *New York Times*, February 16, 2012, www.nytimes.com/2012/02/19/magazine/shopping-habits.html?pagewanted=all.

6. Charles Duhigg, "Campaigns Mine Personal Lives to Get Out Vote," *New York Times*, October 13, 2012, www.nytimes.com/2012/10/14/us/politics/campaigns-mine-personal-lives-to-get-out-vote.html?pagewanted=all.

7. Bill Keller, "Invasion of the Data Snatchers," *New York Times*, January 13, 2013, www.nytimes.com/2013/01/14/opinion/keller-invasion-of-the-data-snatchers.html?pagewanted=all.

8. James Manyika et al., "Big Data: The Next Frontier for Innovation, Competition, and Productivity," McKinsey Global Institute, May 2011, www.mckinsey.com/insights/mgi/research/technology_and_innovation/big_data_the_next_frontier_for_innovation.

9. William Eggers and John O'Leary, *If We Can Put a Man on the Moon: Getting Big Things Done in Government* (Boston: Harvard Business Press, 2009).

10. Everett Rogers, *Diffusion of Innovations*, 5th Edition (New York: Free Press, 2003).

11. Eggers and O'Leary, *If We Can Put a Man on the Moon.*

12. John Kingdon, *Agendas, Alternatives, and Public Policy* (Boston: Little Brown, 1984).

13. Y. Hasenfeld and T. Brock, "Implementation of Social Policy Revisited: A Political Economy Perspective," *Administration & Society* 22 (1991): 451-79.

14. J. Pressman and A. Wildavsky, *Implementation* (Berkeley, CA: University of California Press, 1984).

15. M. Gladwell, *The Tipping Point* (New York: Random House, 1999).

16. Paul Plsek, "Complexity and the Adoption of Innovation in Healthcare," presentation to conference on Accelerating Quality Improvement in Healthcare: Strategies to Speed the Diffusion of Evidence-Based Innovations, Washington, DC, January 27, 2003.

17. R. D. Stacey, *Complex Responsive Processes in Organizations: Learning and Knowledge Creation* (New York: Routledge, 2001).

Part III
Healthcare Analytics Implementation Methods

11

Grasping the Brass Ring to Improve Healthcare Through Analytics: Implementation Methods

Dwight McNeill

The first two parts of the book provided an overview of the challenges, opportunities, and fundamentals for analytics to improve healthcare. The next two parts of the book provide solutions (Part III) and examples of best practices (Part IV). The six chapters in Part III provide state of the art and science solutions to some of the most vexing analytic challenges facing healthcare. These solutions directly address the Healthcare Value Framework to reduce costs, improve outcomes and revenues, and transform the business.

Using the EHR to Achieve Meaningful Results

One of the most important challenges for healthcare analytics is to support healthcare reform through the Affordable Care Act (ACA) in at least three key areas: Insurance reform (especially health insurance exchanges), the Center for Medicare and Medicaid Services (CMS) innovations (especially accountable care organizations (ACOs) and the consumer oriented and operated plans CO-OPS), and health information technology (HIT) (especially meaningful use). All of these areas require advancements in analytics. The most critical barrier to the full expression of analytics is the need to digitize and connect the data "pipes" and integrate new and diverse data. Digitizing the medical record, that is, the electronic health record (EHR), finally took off in 2009 with the Health Information Technology for Economic and

Clinical Health (HITECH) Act. HITECH, through Medicare and Medicaid, provides incentives to physicians and hospitals that adopt and demonstrate "meaningful use" of EHR systems. According to a 2012 National Center for Health Statistics (NCHS) Data Brief, more than 50% of all physicians had adopted an EHR system by the end of 2011, and of the remaining 50%, half plan to purchase or use one already purchased within the next year. Similarly, a 2012 survey of U. S. hospitals indicated that EHR adoption increased from 16% in 2009 to 35% in 2011.[1]

Meaningful use focuses providers on using the EHR information to improve clinical practice, not just to comply with regulations by installing EHR systems. There are many compliance demands of the ACA in addition to meaningful use. Core administrative IT systems need to be ramped-up to provide basic reporting. For many providers, keeping up with the compliance issues consumes most of their analytical time and money. So, it might be hard to see beyond the present demands to an analytics horizon of possibilities.

Full EHR implementation will help in the delivery of care by providing just-in-time information and facilitating coordination among key providers. But after these data are used for the specific clinical purposes, they become digital exhaust and are seldom repurposed. So, the EHR should not become another siloed data bank. The healthcare system needs to move into an integrated information management system that combines the EHR data, and all of its unstructured data challenges, with other person-based data to improve outcomes and reduce costs.

Deborah Bulger and Kathleen Aller, in Chapter 12, "Meaningful Use and the Role of Analytics: Complying with Regulatory Imperatives," make the case that "meaningful use" is much more than a compliance issue and that IT adoption and reporting requirements associated with it form a strong foundation on which to build a new approach to managing care. They state that the meaningful use framework seeks to "revolutionize clinical quality measurement" by making it an automatic, low-cost byproduct of the care process itself. An IT infrastructure that encourages the use of technology is a critical competency to facilitate actionable intelligence across the enterprise and distribute it to stakeholders when and where they need it to make decisions.

Improving the Delivery of Care

Improving the delivery of care to achieve outcomes and efficiencies that take the U.S. healthcare system out of last place when compared to other wealthy nations should be a top priority. The voltage drop between what is known (treatment guidelines) and what is done (actual practice) results in the right care being delivered only 55% of the time. Analytics can and must show the way. Making use of the data by embedding it in clinical decision making, by turning it into useful, accessible, timely, and user-focused information is where the analytic payoff occurs.

Improving clinical decisions through analytics can occur in three ways:

- Shaping care through decision rules. These include rules for care protocols, drug interactions, diagnosis, and order sets, which can be included in EMRs.

- Monitoring and optimizing performance through balanced scorecards and dashboards, which are used for management review and interventions.

- Supporting physicians and care givers with tools for clinical decision making at the point of care.

Real change in healthcare takes place "on the ground" at the physician/patient level. So, changing physician behavior to improve the delivery of care is the challenge for analytics and healthcare leadership. One might reasonably ask, however, "if you build it, will they come?" Physicians do not use information optimally for a variety of reasons. First is the seemingly impossible task to keep track of all the emerging research about diseases and new treatments. Second, it is not always possible for doctors to get the information they need on a given patient because they cannot find it due to the digital "pipes" problem. Third, even if this information were available it might not

be used because physicians are trained in and have a strong preference for intuitive thinking. Many do not have the predilection to sort
through a lot of information and decision maps to make data-driven
decisions. One indicator of the need for more data-driven decision
making in medicine is that diagnostic errors occur about 20% of the
time.[2]

So, the analytics task is to get the information, guidance, and
insights in the hands of physicians, "their way," just-in-time, and in
their preferred delivery mode.

Glenn Gutwillig and Dan Gaines, in Chapter 13, "Advancing
Health Provider Clinical Quality Analytics," develop the case that
clinical quality analytics must move beyond a focus on transaction to
the ability to measure a health provider's compliance to established
clinical standards of care as well as to analyze the relationship between
compliance and clinical outcomes. The authors present their version
of a "next generation clinical quality analytics solution" and a "Clinical
Quality Workbench." They deconstruct clinical protocols into process
components that describe the clinical setting, protocol entry criteria,
diagnostic steps, evaluation criteria, key decision points, the treatment steps, the evaluation criteria as well as related time intervals,
exit criteria, and the needed outcome measurements. They present a
case study on sepsis and demonstrate how the pinpointing of noncompliance and subsequent action can improve quality and reduce costs.

Medical errors continue to be seemingly intractable to improvement. According to the Commonwealth Fund, one in three adult
Americans reported a medical mistake, medication error, or lab error
in the last two years.[3] This is the highest rate among other wealthy
countries that collect the data, and the magnitude of the difference is
that it is almost twice as high as the best performing countries including France, Germany, and the Netherlands.

Dean Sittig and Stephan Kudyba, in Chapter 14, "Improving
Patient Safety Using Clinical Analytics," concentrate on the detection that an error has occurred that is not solely reliant on a clinician's
decision to report it. They discuss the use of "triggers," or automated
algorithms, to identify abnormal patterns in laboratory test results,
clinical workflows, or patient encounters. They describe algorithms
to identify different types of errors including medical diagnosis,

medication administration, and misuse of EHRs. They detail the analytic infrastructure components required for a successful triggers program, including an advanced EHR system, a clinical and administrative data warehouse, a set of clinically tested algorithms or triggers, and a team of clinicians responsible for investigating and managing the incidents identified by the triggers.

Managing the Health of Populations

Managing the health of populations is much different from the medical management of individuals. Medical management in the healthcare system is largely about managing sickness. That's the medical model. Population health is about the production of health, both by preventing illness and by limiting its impact on healthy functioning. In fact, the determinants of health are largely outside the healthcare system. Individual behavior is the strongest predictor of health accounting for 40% of good health, whereas healthcare contributes only 10%. Therefore the approach to producing health is different and includes a raft of other interventions than just what the doctor orders, including social interventions and a reliance on behavior change at the individual/patient/member level.

The incentives for investing in population health are getting better, including changes in payment policy from fee-for-service to global payments, a focus on outcomes and payment accordingly, and changes to the way health is produced. Population health has its roots in public health and has been the province of governments. But with the payment and outcome policy changes under way, including the regulations for health insurance exchanges and reform of insurance practices such as no denial of coverage for preexisting coverage, health plans will need to manage their collective members' health and demonstrate their performance in a transparent way. This will require health plans to take their members seriously because they are their new customers in addition to their stalwarts, employers.

The analytics of population health management are different from clinical management. On the one hand it requires a great deal more knowledge about and the relationships with people to engage them in the coproduction of health. Claims and traditional clinical data are not enough. The various components of health including the World Health Organization's definition physical, mental, and social well-being will require different domains of measurement and at various levels of aggregation, from clinical practices, to healthcare systems, to communities and states. Imagine what would happen if municipalities were held accountable for the health of their citizens just as they are for education, roads, public safety, and jobs.

Stephan Kudyba, Thad Perry, and John Azzolini, in Chapter 15, "Using Advanced Analytics to Take Action for Health Plan Members' Health," detail the difficulty of developing, implementing, and managing population-based care programs. They present a conceptual framework, based on "hot spotting" techniques, that defines the information requirements, analyses, and reporting that will lead to actionable results. They concentrate on the need for proactive predictive analytics that can identify likely future poor health or high cost candidates, who can be optimally impacted by programs, and who will get sufficiently engaged with the care management process to make it a success. They emphasize the foundational need for diverse, robust, integrated, and "clean" data.

Adopting Social Media to Improve Health

Social media is about people using online tools, such as Twitter and Facebook, along with platforms such as mobile, to share content and information. It's the combination of the tool and platform that makes social media so "combustible." The combo is creating a social revolution. On the one hand, it satisfies consumers' long-held wish for convenience, simplicity, immediacy, autonomy, and technology that works for them. On the other hand, it democratizes data by changing the locus of control, emphasizing the power of networks, and how products and services, including healthcare, can be purchased, evaluated, and improved. In healthcare, it could enable more patient/people engagement in the decisions about their personal information and their health.

The implications for healthcare analytics have not totally emerged at this time. Other industries have used social media for marketing, sentiment analysis, and brand management, and healthcare is following their lead. The data for this purpose of analytics have largely come from "scraping" data from digital sources including social media sites and using them for marketing purposes. But, the real potential may lie in gathering much more relevant data from individuals with their consent and engendering their partnership to engage in data sharing activities that help them improve their life.

David Wiggin, in Chapter 16, "Measuring the Impact of Social Media in Healthcare," provides an overview of current and emerging uses of social media specifically for healthcare and proposes an analytical model to measure its impact. He focuses on two general areas of impact, including provider collaboration/education and patient health, including patient education, patient affinity groups, patient monitoring, and care management. He explores how to quantify the value of social media and asserts that its real contribution is in improving population health and that the best source of data may come from patients themselves in the form of surveys. For example, he notes that patient affinity groups collect self-reported data from its members about what works and does not work, which can contribute to comparative effectiveness studies.

Notes

1. E. Jamoom, P. Beatty, A. Bercovitz, et al., "Physician Adoption of Electronic Health Record Systems: United States," 2011. NCHS data brief, no 98. Hyattsville, MD: National Center for Health Statistics, 2012.

2. Pat Croskerry, "Clinical Decision Making and Diagnostic Error," presentation at Risky Business London, May 24, 2012, www.risky-business.com/talk-128-clinical-decision-making-and-diagnostic-error.html.

3. Commonwealth Fund, "Why Not the Best? Results from the National Scorecard on U. S. Heath System Performance," 2011, www.commonwealthfund.org/Publications/Fund-Reports/2011/Oct/Why-Not-the-Best-2011.aspx?page=all.

12

Meaningful Use and the Role of Analytics: Complying with Regulatory Imperatives

Deborah Bulger and Kathleen Aller

The Health Information Technology for Economic and Clinical Health (HITECH) provisions in the 2009 American Recovery and Reinvestment Act (ARRA) created a tremendous opportunity for physicians, hospitals, and health systems to adopt electronic health record (EHR) systems. The legislation includes significant financial incentives designed to accelerate EHR use and ultimately reduce healthcare costs by improving quality, safety, and efficiency. However, the incentives are tied to demonstrating meaningful use of certified EHR technology based on specific measures and milestones that must be documented and reported.

On July 28, 2010, the Department of Health & Human Services published two companion rules finalizing Stage requirements for healthcare providers and for certified technology.[1] Under the final rule, eligible providers and hospitals need to report the results of a core set of measures and a menu set of measures as part of the demonstration process. These measures are paired with meaningful use objectives—such as using computerized provider order entry (CPOE) or recording smoking status—and apply to eligible providers, hospitals, or both. One of the core objectives is to report automatically computed quality measures to the Centers for Medicare and Medicaid Services (CMS). Within this one objective are 15 clinical quality measures for hospitals, evaluated for all patients regardless of payer. The same objective for eligible providers breaks down further into core and specialty measures, but there are many fewer for a given provider.

Since the ARRA legislation became law, there has been a flurry of activity, including the federal rule-making process for meaningful use and certification. Eligible providers and hospitals that plan to qualify for incentives must demonstrate meaningful use; health information technology (IT) vendors are responsible for achieving EHR certification. Vendors responded rapidly to the requirements to optimize EHR objectives and measurement. At this writing, the Office of National Coordination for Health Information Technology (ONC) lists 160 inpatient and 363 ambulatory applications from various vendors that are certified as either complete or modular EHRs for Stage 1 objectives.[2]

To receive full EHR incentives, hospitals must be meaningful users of certified technology by Federal Fiscal Year 2013. Eligible hospitals and providers are investing significant financial and human resources in assessing the current status of EHR technology in their organizations and devising strategies to mitigate any potential gaps. A report published by Accenture in January 2011 estimates that 90% of U.S. hospitals will need to install or upgrade EHR technology during the next three years and that approximately 50% are at risk of not achieving meaningful use by 2015 when penalties will begin to take effect.[3] A recent HIMSS Analytics survey reports that only 27% of hospitals responding to the survey since the final rules were published expect to meet Stage 1 meaningful use requirements by July 2012.[4]

The findings from these and other studies suggest that achieving meaningful use will require an accelerated investment in technology but will also demand a strategic commitment to the use of information to drive behaviors and ensure EHR adoption across the enterprise. This convergence of technology, information, and behaviors is a critical predictor of successful business performance as described by Marchand, et al.[5] In this regard, "meaningful use" may be one of the most influential developments in recent history to drive analytics maturity in the healthcare market.

Supporting EHR Adoption with Analytics

A basic HITECH assumption is that measuring the quality of clinical care and IT usage should flow automatically from using an EHR system during patient care. For this automation to occur, several prerequisites must be in place:

- Certified EHR functionality must be deployed throughout the provider organization. An example of IT functionality is nursing documentation.

- That functionality must include the necessary clinical content to support the required data collection. For example, to record the smoking status of a patient, a single structured documentation element needs to be in place.

- The functionality must be deployed using a prescribed workflow and methodology to help ensure that the data collected are consistent and comprehensive. To build on the previous example, the prescribed workflow would embed collecting smoking status for all patients age 13 and older in the admission assessment conducted by a caregiver, and it would cue the caregiver that documentation is missing or unconfirmed until properly collected or updated.

Within the framework of the meaningful use rule, healthcare providers must achieve and report specified results for each of the IT functionality measures associated with meaningful use objectives. In the previous example, incentive payments depend on documenting specific findings for more than 50% of the applicable patients treated during the measurement period. No organization wants to reach the end of a reporting period only to tally up its results and find them deficient. It is therefore essential that EHR functionality be supplemented with a way to continuously measure the adoption of key EHR components and track internal performance against the full set of IT measures. Certified EHRs must be able to calculate metrics associated with the objectives to quantify capabilities and adoption levels. The intent is to record whether:

- Features are activated
- Communication functionality has been tested
- EHR components are in use by a percentage of users
- Basic data of interest are being collected for a portion of the patient population

To be effective, information must reach the people charged with improving performance in a timely and appropriate fashion. For measures related to patient care activities, reporting must reach caregivers in real-time. For this reason, there are meaningful use objectives to implement drug and allergy checking, and to align that checking with CPOE-based medication orders. By regularly monitoring the measure across individual units, caregivers, or shifts, the organization can address issues of adoption or care processes that affect patient care outcomes. For measures that affect payments such as meaningful use incentives, managers can ensure that performance does not drop below required levels. This aggregated, trended reporting, supplied through an organization's enterprise intelligence solution, will increase visibility to meaningful use objectives across all stakeholders.

A well-designed enterprise dashboard should support the analysis of multiple aspects of the meaningful use objectives. For example, it is not enough to simply track the percentage of orders placed by authorized providers using CPOE. It is also necessary to know who is ordering, what they are ordering, and more importantly, who is not using CPOE. This information is essential to put in place the necessary coaching, training, and system modifications to support adoption, and the attainment of meaningful use. As an organization progresses along the adoption path, it may decide to set a more aggressive goal than the Stage 1 meaningful use requirement for CPOE.

Shifting the Quality Analysis Paradigm

Clinical quality measurement is already a prominent component of The Joint Commission certification process and CMS incentive payment mechanisms such as Hospital Inpatient Quality Reporting Program (Hospital IQR). However, the computation and submission of measures today is typically a cumbersome, costly, manual process that is almost entirely retrospective, while doing little to drive behavior change. Several organizations have tried to automate the data collection with varying degrees of success. Mapping the data elements needed to calculate the current measures to the technology required to capture them suggests that, assuming a completely digital environment, approximately 60% of the data could be captured electronically but that over half of those elements would require human intervention for validation and quality control. Primary barriers to automated data collection fall into three categories:

- Measure specifications that are incompatible with automated data capture and that specifically require human intervention.

- Lack of standardized documentation across the healthcare system. As an example, smoking status may be documented by different caregivers in multiple locations using a variety of text fields and lookup tables.

- Low adoption rates of technology from which key data must be extracted. Gartner reports that only 20% of hospitals use clinical documentation in the emergency department, while only 5% use perioperative charting and anesthesia documentation as part of an EHR.[6]

Besides the barriers to data access, the descriptive nature of the reporting has prevented organizations from developing a competency for analytics much beyond ad hoc reports and query drilldown. Furthermore, the intense focus on regulatory requirements relegates the healthcare constituent to a secondary audience with access to results only after they have been submitted to CMS months after the patient has been discharged (see Figure 12.1).

The meaningful use framework seeks to revolutionize clinical quality measurement by making it an automatic, low-cost by-product of the care process itself. Conceptually, this goal assumes that the required data for measure calculation are captured within the EHR during care and then flow seamlessly to reporting and data submission mechanisms. To accomplish this, existing measure specifications must be completely redesigned, or retooled, to transform them from manual measures to so-called eMeasures.

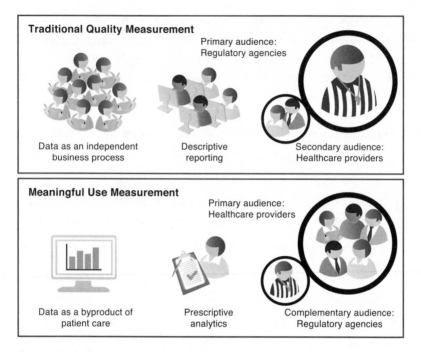

Traditional Quality Measurement

Primary audience:
Regulatory agencies

Data as an independent
business process

Descriptive
reporting

Secondary audience:
Healthcare providers

Meaningful Use Measurement

Primary audience:
Healthcare providers

Data as a byproduct of
patient care

Prescriptive
analytics

Complementary audience:
Regulatory agencies

Figure 12.1 Changing the quality reporting paradigm

The initial work on measure redesign was done by the Health Information Technology and Standards Panel (HITSP), under contract to CMS, to modify three sets of existing quality measures to use EHR-generated data directly. Fifteen of the original 16 measures, covering Emergency Department (ED) throughput, stroke, and venous thromboembolism (VTE), were adopted in the final rule for use by eligible hospitals.

Managing quality measurement under this new framework represents a significant paradigm shift for healthcare organizations. Historically, quality measurement has been treated as a *separate business process*, employing a host of abstractors to collect data after the fact. The new paradigm treats data as a *core clinical competency* driving information to and from the point of care. Using data as a by-product of patient care will eventually reduce dependency on manual chart abstraction but will also demand new analytics skills on the part of clinicians. While *extrinsic drivers* may define quality measurement in this new model, organizational stakeholders clearly become the primary audience for the results, creating *intrinsic motivation* for organizational improvement. Although measure calculation will continue to be retrospective, there should be less lag time for reporting with electronically generated measures. Additionally, the measure design begins to lay the foundation for more prescriptive clinical workflows, and hence more prescriptive analysis—that is, the ability to use evidence-based guidelines and subsequently capture data that facilitate better outcomes and predictive analysis. This approach drives hospitals up the analytics maturity curve by enabling use of real-time alerts and query tools to ensure that more uniform care is delivered in a timely manner.

In Stage 1, CMS has not specified achievement targets for clinical quality measures. However, the measures lend themselves to statistical analysis and forecasting aspects of the care process. For example, measures of ED admit to departure times facilitate the use of statistical process control tools to evaluate variation, correlate root cause, forecast capacity, and anticipate throughput barriers. By measuring incidence of preventable venous thromboembolism (VTE), the VTE measure set strongly correlates process of care (i.e., adherence to evidence-based guidelines for VTE prevention), with patient outcome. As a result, organizations can capitalize on the propensity of these measures to change behaviors and drive improvement.

Driving Analytics Behaviors

CMS has clearly adopted the principle of defining measurable goals and aligning them with incentives. It now lies with participating eligible hospitals to manage performance against those goals. Given the specificity of those goals, one may question whether meaningful use will help or hinder an enterprise journey toward analytics maturity. To answer that question, we organized measures along three aspects of care:

- Primary stakeholder accountability
- Action required to achieve the measures
- Method by which the action is taken

By combining the objectives-based and quality measures in this manner, patterns emerge that enable targeted coaching and change management.

Accountability

There has been considerable focus on CPOE implementation and adoption in hospitals, with the physician as the primary "owner" for approximately 30% of the combined IT functionality and quality measures (see Figure 12.2)—for example, the use of CPOE for medication orders or the appropriate discharge medication orders. Sixteen percent of the measures are primarily directed at nursing care, such as charting vital signs or providing stroke education. However, 34% of the measures are based on care that is typically interdisciplinary, such as maintaining up-to-date problem lists or the occurrence of preventable VTE. This suggests a need for a cross-functional approach to improvement and shared accountability for the results. Organizing measures along the lines of accountability and using analytics to create transparency empowers stakeholders to take appropriate action to ensure continued success.

Primary Responsibility for the Objective or Measure

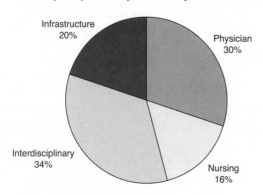

Figure 12.2 Accountability for measures

Action

Viewed another way, when one considers the action (see Figure 12.3) that the measure describes, 28% of the measures evaluate activities that occur during transitions of care, such as medication reconciliation or ability to provide a summary of care at discharge. Care coordination—"insuring that patients receive well-coordinated care within and across all healthcare organizations, settings, and levels of care"—is one of eight key objectives developed by the National Priorities Partnership. Aimed at improving quality and reducing disparities in care, it will likely be a continued focus of meaningful use in Stage 2 and 3. While compliance is measured at the individual objective level, it is important to view measures holistically. The ability to analyze disparities in care depends on the availability of descriptive data (such as race, ethnicity, language need, and socioeconomic status) for populations at risk for poor quality care.[7] Therefore, objectives such as recording patient demographics play a significant role in monitoring potential disparities in care in underserved populations. In other words, failure to achieve meaningful use in one objective may have a direct impact on another.

Action That Defines the Objective or Measure

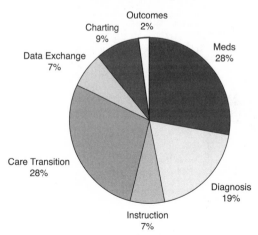

Figure 12.3 Action provided by accountable stakeholder

Method

How the objectives are achieved represents the most complex aspect of measure management (see Figure 12.4). Impressively, 23% of the objectives should be achieved electronically, such as the incorporation of lab results or the e-exchange of key clinical information at care transition points, and are a primary function of the IT infrastructure. Objectives such as ensuring overlap thrombolytic therapy for stroke patients requires intervention by multiple stakeholders—the physician, clinical pharmacist, and patient. Nearly 40% of the objectives are achieved through either assessment of the patient or decision points—that is, drug formulary checks or decision to admit from the ED. EHR technology augments and enables care decisions but does not replace critical thinking and clinical expertise. Analytics provides access to knowledge about the care delivery model and the ability to dissect patterns and trends that may impede progress. For example, is a delay in the decision to admit a patient from the ED to an inpatient bed a function of communication, knowledge, or capacity? Each root cause commands a different response. Understanding the "how" of measure compliance combined with the "who" and "what" enables an organization to manage improvement with laser focus while ensuring overall enterprise success.

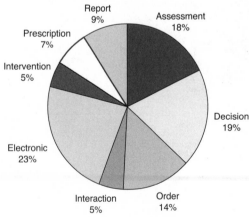

Figure 12.4 Method by which action is taken

"Wringing every drop of value"[8] from meaningful use measures in Stage 1 will lay the foundation for analytics expertise as these measures become more pervasive. As stated by Michael Davis, executive director of HIMSS Analytics, "Hospitals that survive the upcoming healthcare delivery transformation will be organizations that understand the need to use EMRs to collect, manage, share, and analyze data with the intent to continually improve their care delivery processes using best practices and evidence-based medicine protocols."[9]

Anticipating the Future

The development of new measures will likely require longitudinal data to measure across care settings and use data outside the traditional EHR. The National Quality Forum (NQF) has taken over guidance for developing new measures and retooling existing ones as eMeasures. As its work begins to stretch the boundaries of current analytics capabilities, healthcare will likely need tools that are neither widely used nor available today. Both the National Priorities Partnership and ONC's Health IT Policy committee recommend that future stages of ARRA and other federal performance measurement efforts include the measurement of clinical outcomes, safety, and efficiency, necessitating expanded documentation with a clinical decision support system (CDSS). A HIMSS Analytics 2009 survey reports that the adoption rate of CDSS at the physician documentation level is approximately 11%.[10] In Stage 1, physicians must attest to the use of one clinical decision support rule per patient during the reporting period, in preparation for more extensive use. Additionally, as the complexity of measures increases, the need to aggregate disparate sources of data from across multiple care settings will accelerate the need for integrated clinical and financial business intelligence and data warehouse solutions. Gartner research published in July 2010 suggests that less than 5% of hospitals have deployed integrated clinical and financial business intelligence and data warehousing solutions, indicating a need for rapid growth in this emerging market.[11]

While the achievement of meaningful use goals is designed to produce tangible patient care results and significant financial incentives, measurement of IT adoption and related quality improvement represents only one aspect of a broader management imperative. The concept of "aspirational" metrics—that is, aspiring to a new standard—requires that organizations strive to not only meet current requirements but to "move the bar" toward organizational excellence. Sustainable success derives from a multifaceted strategy that requires constant care and feeding, and the management of an overwhelming amount of new data. While pursuing meaningful use incentives, organizations must continue to manage the overall business of healthcare, including

- Demonstrating the competency of the clinical staff
- Ensuring that patient care occurs in a timely, efficient manner in the right care setting
- Recruiting and retaining qualified employees
- Modeling revenues and managing the costs of operations to remain financially viable beyond incentives
- Increasing market share

These business dimensions cannot be measured in silos. The correlation of cost, quality, care coordination, and efficiency is necessary to influence organizational decision making and to remove departmental and political barriers to drive transparency to key business processes.

Conclusion

Meaningful use, if embraced as a foundational catalyst and not a regulatory burden, could well be the driver that promotes analytics maturity in healthcare. The meaningful use and IT adoption and reporting requirements form a strong foundation on which to build a new approach to managing the business of care. Building an IT infrastructure that encourages the use of technology and provides more time to care for patients is the first step. It is also a critical factor in the evolution of information management: the ability to transform data created as a by-product of patient care into actionable intelligence across the enterprise and distribute them to stakeholders when and where they need them to make decisions. This process requires aggregating data from multiple sources and applying healthcare logic and rules. Finally, managing behaviors by instilling the value of information in the people who manage care begins to create a culture of shared accountability for the organization's results.[12]

Notes

1. 42 CFR Parts 412, 413, 422 et al. Medicare and Medicaid Programs; Electronic Health Record Incentive Program; Final Rule 42 CFR Parts 412, et al. Medicare and Medicaid Programs; Electronic Health Record Incentive Program; Proposed Rule, *Federal Register*, Vol. 754, No. 1448, 1/137/28/2010, pp. 443131843-201144588, http://edocket.access.gpo.gov/2010/pdf/2010-17207.pdf.

   45 CFR Part 170 Health Information Technology: Initial Set of Standards, Implementation Specifications, and Certification Criteria for Electronic Health Record Technology; Final Rule *Federal Register*, Vol. 75, No. 144, 7/28/2010, pp. 44589-44654, http://edocket.access.gpo.gov/2010/pdf/2010-17210.pdf.

2. Details may be found at http://onc-chpl.force.com/ehrcert/CHPLHome.

3. "Lessons from EHR Pros," HealthData Management, http://www.healthdatamanagement.com/news/ehr-implementation-lessons-meaningful-use-41761-1.html.

4. Read more: EMR Daily News, http://emrdailynews.com/2011/02/24/new-himss-analytics-data-shows-44-of-hospitals-likely-to-be-ready-for-stage-1-of-meaningful-use/#ixzz1HoQwrlz7.

5. Adapted from D. A. Marchand, W. J. Kettinger, and J. D. Rollins, "Information Orientation: The Link to Business Performance" (Oxford: Oxford University Press, 2000).

6. Hype Cycle for Healthcare Provider Applications and Systems, 2010, Thomas J. Handler, M. D., July 27, 2010 ID number G00205364.

7. "Future Directions for the National Healthcare Quality and Disparities Reports," Institute of Medicine Report Brief, April 2010, www.iom.edu/ahrqhealthcarereports.

8. Thomas Davenport et al., *Competing on Analytics: The New Science of Winning* (Harvard Business School, 2007).

9. Michael W. Davis, Executive Director, HIMSS Analytics, "The State of U.S. Hospitals Relative to Achieving Meaningful Use Measurement."

10. Ibid.

11. Hype Cycle for Healthcare Provider Applications and Systems, 2010, Thomas J. Handler, M. D., July 27, 2010, ID number G00205364.

12. Adapted from D. A. Marchand, W. J. Kettinger, and J. D. Rollins, *Information Orientation: The Link to Business Performance* (Oxford: Oxford University Press, 2000).

13

Advancing Health Provider Clinical Quality Analytics

Glenn Gutwillig and Dan Gaines

Current systems typically measure healthcare as a series of transactions rather than a process that may or may not be executed as designed. To reach the next level of quality in care delivery, practitioners require a comprehensive set of facts around healthcare delivery including compliance with every aspect of the protocol and the outcome(s) of care.

The public and private payer focus on managing population health is driving providers to transition from focusing on acute care delivery to an outcome based model. This focus is designed to reward delivery efficiency, lower cost, and accountability across the full continuum of care providers.

Increasingly educated and vocal consumers are demanding platforms that will enhance their participation in the healthcare continuum. They are demanding proof of quality care and positive outcomes in addition to reasonable access and cost. These metrics require that data typically imbedded in electronic medical records (EMRs) and across health information exchanges (HIEs) be available to the consumer.

Clinical Quality Analytics Background

Over the past 25 years, provider organizations have attempted to adopt evidence-based clinical treatment guidelines as a cornerstone of their clinical quality programs. Their clinical quality measures are based on specific evidence-based practices, developed over time, which have been shown to provide the best care results to the most people. Example areas where clinical quality is typically measured and reported include heart attack, heart failure, stroke, pneumonia, and surgical care protocols and processes. Hospitals all across the United States measure and report their performance quality in these clinical areas. Today, data are generally updated and trended on a quarterly basis. Each measure represents what percentage of the time patients received the recommended care, so a higher score represents better performance. In most organizations, these clinical guidelines are implemented through education and, in some cases, EMR clinical workflows and order-sets.

To address the ongoing need to improve quality and meet regulatory benchmarks, most if not all hospital organizations have deployed a clinical quality organization. At their core, these organizations are chartered with the responsibility to measure the quality of care being delivered and work with the clinical leadership to identify improvement opportunities and drive process change. The assessment of adherence to defined clinical guidelines and processes has been at the heart of measuring the quality of the care process. However, compliance with the guidelines continues to be both challenging and problematic for most providers. For example, the New England Health Institute's (NEHI) research points to several contributing factors to poor compliance, including counterincentives in the payment system, the lack of integrated data, physician culture because many doctors receive little feedback on adherence to evidence-based clinical practice guidelines, and the development of guidelines themselves. In particular, the research points to the lack of inclusion and transparency in current guideline development leading to a decreased level of trust by physicians in the clinical practices they are being asked to follow.

Moving forward, the ability to measure a health provider's compliance to process as well as to analyze the relationship between process compliance and population clinical outcomes will be a fundamental requirement. In addition, evaluating outcomes and costs associated with care presents an additional opportunity to improve care delivery and providers' management processes.

Collecting, analyzing, and reporting clinical quality data are often difficult and are viewed by many provider organizations as burdensome. In addition, the financial incentives from the payers to focus attention on clinical quality in the health provider community have historically been lacking. There have been incentive programs of various forms, but they have been, for the most part, inconsistent and of minimal incremental value to the providers.

Today, the drivers for improved quality analytics are intensifying, and the nature of analytics is evolving. Key drivers include

- **Reimbursement model changes**—Pay for Performance (P4P), Accountable Care Organizations (ACOs), and value-based purchasing will demand greater focus on quality, with revenue and profits at risk. Many health systems now share the view that the path to becoming a financially successful ACO must include constant monitoring of clinical performance, evaluation and improvement of clinical quality, reliability, and operational efficiency over the full range of providers from community-based to tertiary medical centers. It is becoming increasingly clear that payers of all types will demand greater focus on quality, with increasing amounts of revenue at risk for noncompliance.

- **Government reporting requirements**—Never events, meaningful use, Center for Medicare/Medicaid Services (CMS) Core, The Joint Commission (TJC), formerly the Joint Commission on Accreditation of Healthcare Organizations (JCAHO)—requirements are all evolving toward outcomes management.

- **Evolving quality measurement programs**—The Physician Quality Reporting Initiative (PQRI) approach is phasing out in favor of more stringent National Quality Forum (NQF) standards. Outcomes will become more important than process compliance.

- **Evidence based care**—A significant increase in clinical protocols will require an in-depth understanding of compliance and process management. In evidence-based medicine (EBM) the ability to analyze the impact of specific interventions is becoming a key issue.

Accenture's Health Analytics team working with leading providers is pioneering the design, development, and deployment of next generation clinical quality analytics. These solutions support health providers' quality improvement initiatives and will capture clinical treatment guidelines as process definitions, evaluate and analyze clinical encounters for their compliance to the appropriate guidelines, and assess the impact of noncompliance on clinical and financial outcomes. Identifying noncompliance issues will facilitate the understanding of the root cause of compliance issues allowing for more focused and efficient process improvement programs resulting in both improved outcomes and reduced costs. Finally, this approach creates the ability to track and analyze the results of improvement programs more efficiently and with greater accuracy than is typically possible today through either standard outcomes or key performance indicators (KPIs) monitoring.

Access to EMR data is an essential prerequisite, but not the complete source of data. The data used to measure compliance to protocols must be structured and augmented to act as a meaningful source for clinical quality information. Currently EMR systems are focused on patient centric "record keeping" and used to manage care at the individual level.

While timeline data is captured in EMR systems, the data are not organized to analyze the temporal (chronological) process of care. The next generation of clinical quality analytics will provide the ability to

- Focus on "events" at both the patient and specific target population levels.
- Analyze a treatment outcome from a quality perspective, either a relatively short-term event like sepsis, or long-term treatment of a chronic condition like diabetes.
- Compare actual events to expected or desired events, down to the patient level, for specific populations under care.
- Derive and predict the impact of compliance.

Next Generation Clinical Quality Analytics Solutions

To dramatically improve the state of clinical quality analytics several key capabilities and technologies are integrated to create a comprehensive solution:

- The ability to model multistage complex clinical processes using a time based clinical process modeler for patients with a range of demographic and clinical factors
- The ability to compare enriched data sets of actual care events against the yardstick of the clinical protocol process maps
- The availability of text mining capabilities and a user interface for required manual chart reviews
- A compliance evaluation and analytics engine to assist with root cause analysis and provide predictive capabilities that can model the impact of performance improvement
- A clinical review and collaboration environment
- An analysis and visualization environment for clinical quality data
- A workbench model as shown in Figure 13.1 to support numerous parallel quality improvement initiatives

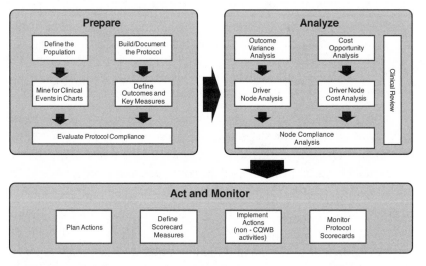

Figure 13.1 Accenture's Clinical Quality Workbench Analysis Architecture (Source: Accenture, © 2011)

The Clinical Quality Analytics in Action— Case Study: Sepsis Overview

To test the potential for this new set of quality metrics, several case studies have been initiated. Though not very common, sepsis is a serious medical issue in hospitals, arising from infections that overwhelm the bodies' defenses and can result in death in as many as 40% of cases. Sepsis is characterized by a whole-body inflammatory state (called a systemic inflammatory response syndrome or SIRS) and the presence of a known or suspected infection. The body may develop this inflammatory response by the immune system to microbes in the blood, urine, lungs, skin, or other tissues. A lay term for sepsis is blood poisoning, more aptly applied to septicemia, as used in the example. Severe sepsis is the systemic inflammatory response with infection, plus the presence of organ dysfunction.

Sepsis was selected as an example to demonstrate the Clinical Quality Workbench for several reasons:

- Treatment of sepsis has been identified as a potential cost and outcome opportunity for leading health providers.
- A standing protocol to manage and treat the condition is available.
- Outcomes and progression can be clearly defined.
- Most of the data necessary to assess compliance should be available in the hospital's EMR system.

To facilitate robust analytics capabilities we defined the existing clinical protocol for sepsis as a time-based process.

This process definition describes the clinical setting, protocol entry criteria, diagnostic steps, evaluation criteria, key decision points, treatment steps, and evaluation criteria as well as related time intervals, exit criteria, and the needed outcome measurements.

In addition, while much of the clinical protocol data is available in discreet data elements, such as lab, pharmacy, and vital signs, some critical data points can often only be found in the free text areas of the EMR, such as physician and nursing notes and radiology reports. Thus within the context of each clinical protocol, it is often highly desirable to capture valuable unstructured or semistructured data (related to notes, images, etc.) and combine it with the structured EMR data to perform a holistic and robust analysis.

Figure 13.2 highlights a specific process component of the sepsis protocol and illustrates how an actual encounter is compared to the protocol. Each process component defines specific functions, tasks, and related measurement criteria including the definition of compliance for the step. For example, in Step B14 the mental status of the patient must be checked and documented every 2 hours. In this single process step, a high rate of noncompliance, (49%), was observed, which directly correlated to increased mortality rates.

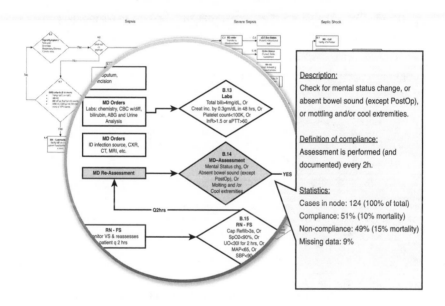

Figure 13.2 Clinical protocol for sepsis (Source: Accenture, © 2011)

The output of the clinical quality analysis correlated advanced sepsis and septic shock with low compliance scores for one unit in the study hospital (see Figure 13.3).

The data supported further investigation of process compliance patterns for multiple delivery units within the hospital and root cause analysis. Figure 13.4 depicts a more detailed analysis of compliance to protocol by unit 2.

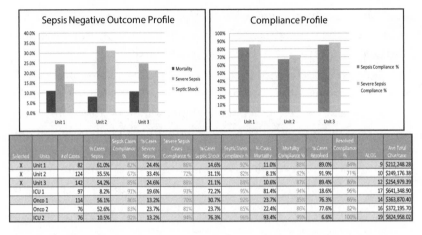

Selected	Units	# of Cases	% Cases Sepsis	Sepsis Cases Compliance %	% Cases Severe Sepsis	Severe Sepsis Cases Compliance %	% Cases Septic Shock	Septic Shock Compliance %	% Cases Mortality	Mortality Compliance %	% Cases Resolved	Resolved Compliance %	ALOS	Ave Total Chart/case
X	Unit 1	82	61.0%	82%	24.4%	86%	14.6%	92%	11.0%	88%	89.0%	84%	9	$212,248.28
X	Unit 2	124	35.5%	67%	33.4%	72%	31.1%	82%	8.1%	82%	91.9%	71%	10	$249,176.38
X	Unit 3	142	54.2%	85%	24.6%	88%	21.1%	88%	10.6%	87%	89.4%	86%	12	$254,979.39
	ICU 1	97	8.2%	91%	19.6%	93%	72.2%	95%	81.4%	94%	18.6%	96%	17	$641,348.90
	Onco 1	114	56.1%	80%	13.2%	70%	30.7%	92%	23.7%	85%	76.3%	86%	14	$363,870.40
	Onco 2	76	52.6%	83%	23.7%	81%	23.7%	85%	22.4%	86%	77.6%	82%	16	$372,195.70
	ICU 2	76	10.5%	92%	13.2%	94%	76.3%	96%	93.4%	95%	6.6%	100%	19	$824,958.02

Figure 13.3 Correlation between advanced sepsis/septic shock and low compliance scores (Source: Accenture Health Analytics Research, © 2011)

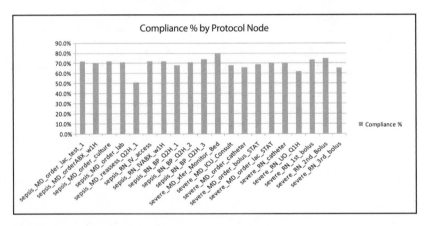

Figure 13.4 Compliance to protocol by unit 2 (Source: Accenture Health Analytics Research, © 2011)

Figure 13.5 illustrates significant variations in compliance to protocols by shift staff and specifically variation to protocols around the time of shift change. The contributing factors were identified, and feedback and process improvement were developed to support corrective efforts.

Figure 13.5 Compliance variations by staff by shift and by hour (Source: Accenture Health Analytics Research, © 2011)

The ability to pinpoint areas of noncompliance and to improve quality will drive significant clinical and financial impact for healthcare providers. For this one protocol:

- The impact of improving compliance on unit 2 to just the average of the hospital's two other units:
  - Would be expected to result in four fewer deaths from sepsis on this unit alone per year
  - Would be predicted to result in 23% fewer patients progressing to severe sepsis or septic shock, saving the hospital $250k to $500k overall in this unit
- Lifting performance across units for this one area of sepsis could reduce cost by $1.5 to $2 million overall.

Extending this type of clinical quality analysis across the hundreds of protocols in place in a typical hospital can drive significant impact. For example, by extrapolating across a series of protocols for a medium to large hospital system, the improved cost performance opportunities are in the $50+ million per year range.

Conclusions

The pressure to improve quality across the healthcare continuum has continued to increase and is being driven by patients as well as regulations. New technology has provided an opportunity to make significant improvements in care through understanding the detailed sources of variation and error in the care delivery system. Accenture is working with clients to bring this new technology to bear and to develop the models that show significant promise in helping to make significant improvements in the understanding of healthcare quality issues. These capabilities will enable providers to understand and achieve optimal outcomes, enhance their quality processes, and develop an organization-wide awareness of quality and its impact on the both the lives under their care and the financial bottom line.

14

Improving Patient Safety Using Clinical Analytics

Dean F. Sittig and Stephan Kudyba

Clinical analytics, coupled with the data extracted from real-time, point-of-care electronic health record systems, has the potential to significantly increase patient safety through the use of "triggers" or computer-based algorithms that automatically identify patterns in data that suggest errors have occurred in healthcare-related activities. This chapter describes research currently being conducted to identify three different types of errors in 1) medical diagnosis, 2) medication administration, and 3) use of electronic health records (EHRs). Using these new analytic measurement systems, researchers can better evaluate the effect of the myriad health information technology interventions on patient safety.

Introduction

Patient safety refers to the need for those involved in the delivery of healthcare to avoid harming the patients they are treating.[1] Beginning with the Institute of Medicine's landmark 1999 report titled "To Err is Human," there has been an increasing emphasis on all aspects of medical errors, where medical error is defined as "the failure of a planned action to be completed as intended or the use of a wrong plan to achieve an aim."[2] Even with this increased emphasis, there has been no reduction in the number or severity of medical errors as reported by clinicians or as identified after the fact by clinicians trained to read through a patient's medical record looking for evidence of error.[3] These errors can occur in almost any aspect of the clinical work process, including diagnosis (e.g., creating an incomplete or incorrect list of potential diagnoses, ordering the wrong diagnostic tests, or incorrectly interpreting the test results), treatment planning (e.g., selecting the wrong therapeutic procedure) or administration (e.g., performing the wrong therapeutic procedure or performing the correct procedure incorrectly), and in failure to follow up on known clinical problems (e.g., failure to perform a needle biopsy on a patient with an abnormal mammogram).

Before an individual or organization can begin fixing the problems that lead to these errors, they must be able to detect that an error has occurred and investigate its underlying causes. Without significant progress in enhancing the understanding of the causes of medical errors, and reducing their number and severity, patients will continue to suffer from unnecessary patient harm as a result of erroneous or delayed diagnosis, incorrect or unnecessary medication administration or therapeutic procedures, for example.

Recently patient safety researchers have demonstrated that simply relying on clinicians to self-report these errors overlooks the vast majority of occurrences. Toward this end, researchers have begun to develop "triggers" or automated algorithms to identify abnormal patterns in laboratory test results, clinical workflows, or patient encounters. These triggers are easily computed using existing clinical analytic techniques applied to large clinical and administrative data warehouses.

Background

Before an organization can develop an automated error detection system based on automated triggers, several key clinical analytic infrastructure components are required:

- An advanced electronic health record system (EHR) that captures information on all patient visits, clinical problems, medications, laboratory test results, and therapeutic procedures using a controlled clinical vocabulary. Simply recording all this information in a free-text note precludes any automated analysis without significant natural language processing (NLP) to identify specific data items.

- An offline clinical and administrative data warehouse that re-indexes the patients' data to allow rapid retrieval of longitudinal patient information (e.g., find all patients with a cholesterol level greater than 240 milligrams per deciliter of blood) rather than the patient-centered view (e.g., what is patient X's cholesterol level) that is required by clinicians to manage an individual patient. This offline data warehouse should also incorporate and integrate data from multiple external data sources (e.g., billing data, pharmacy dispensing data, geographic location, etc.) to allow researchers to better identify and isolate the patient's clinical context.

- A set of clinically-tested algorithms or "triggers," implemented as queries against the data warehouse, that can identify specific patterns that have a high probability of association with a medical error (e.g., an order for an antidote for example, Naloxone) to a medication overdose (e.g., morphine).[4] These algorithms need to be carefully tested to ensure that they have suitable sensitivity (i.e., identify all the important cases) and specificity (i.e., do not identify too many incorrect cases).

- A dedicated team of clinicians responsible for investigating all the "incidents" identified by the triggers. This team of clinicians must be able to quickly ascertain whether the identified individuals truly had the suggested condition and follow up on these findings in an appropriate manner.

Once an organization has all of these technical and social systems in place, they can begin to identify, investigate, report, and manage at least a subset of their clinical errors. The following sections describe three specific examples that illustrate the potential use of clinical analytics to identify various types of clinical errors.

Triggers for Diagnostic Errors

This diagnostic process begins with the initial data gathering phase in which the patient's signs and symptoms are collected via interview, observation, and physical examination. It continues with the hypothesis generation phase in which the clinician compares the patient's signs and symptoms to the medical knowledge base. These hypotheses are then tested in the diagnostic test ordering phase in which the clinician attempts to either confirm or rule out specific diagnoses through the judicious selection of tests of various physiologic processes (e.g., testing the patient's ability to form a clot [necessary before one can safely perform surgery] by ordering coagulation tests—Prothrombin Time, Partial Thromboplastin Time, and International Normalized Ratio). The results of these tests help the clinician make the final diagnosis, which forms the basis for the various treatment options she must consider.

Diagnostic errors (i.e., missed, delayed, or wrong diagnosis[5]) can occur in any clinical setting and are arguably the leading type of error in primary care. Ambulatory practice-related diagnostic errors account for up to 40% of malpractice claims and result in average claims of $300,000.[6] There are multitudes of ways in which these errors can occur, but the end result of these errors is that the patient does not regain his health and often returns to the health system a second or third time seeking help.

In an attempt to identify potential diagnostic errors in the ambulatory setting, regardless of their cause or position in the diagnostic process (e.g., faulty data collection, inadequate medical knowledge, erroneous or lost test results, etc.), Singh et al. developed Structured Query Language (SQL)-based queries to detect the presence of one of two mutually exclusive events: trigger 1) a primary care visit (index visit) followed by a hospitalization in the next ten days; or trigger 2) an index visit followed by one or more primary care, urgent care, or emergency department visits within ten days.[7] Follow-up evaluation of these triggers resulted in a positive error identification rate of 16.1% and 9.4% for triggers 1 and 2, respectively, with an error rate of 4% in cases that met neither screening criterion. Further investigation into the causes of these errors showed that the most common errors were

- Failure or delay in asking the appropriate questions of the patient (e.g., asking a patient with suspected tuberculosis whether she has recently traveled to Asia or sub-Saharan Africa) or ordering the appropriate diagnostic tests (e.g., x-ray or blood analysis)

- Misinterpretation or suboptimal weighting of critical pieces of data from the history and physical examination
- Failure to recognize urgency of illness or its complications.

Similar triggers have been used for diagnostic errors that occur in the hospital or its Emergency Department (ED), for example, patients who return to the ED within 72 hours of hospital discharge[8] or any patient who spends greater than 6 hours in the ED, with similar accuracy in error identification.

Triggers for Medication Errors

Medication administration-related errors are common in all clinical settings as well. Medication errors can occur due to incorrect medications, doses, routes, time of administration, and even administering medications to the wrong patient. In an attempt to develop a tool to help organizations begin to identify these errors, the Institute for Healthcare Improvement (IHI) developed a global trigger tool (GTT).[9] Briefly, the GTT provides a standard methodology for reviewing patient records for triggers, or indicators, of potential adverse events. Each of these triggers is then reviewed by a clinician for final determination of whether an error actually occurred.

Assuming that an organization is using a computerized order entry system, the automatic detection of specific adverse events related to medication administration can be relatively straightforward using the GTT. Example triggers from this tool include looking for abrupt, unanticipated medication stop orders, abnormal laboratory results (e.g., Partial Thromboplastin Time (PTT) greater than 100 seconds in patients taking heparin), or use of an antidote medication (e.g., Diphenhydramine [Benadryl] administration). One can reliably identify errors with a sensitivity approaching 95% and a specificity approaching 100%.

Several hospitals have now automated the entire GTT and on a daily basis scan their entire hospital looking for potential adverse events.[10, 11] Using a list of all potential medication errors, a pharmacist reviews the medical records of all patients to confirm the errors. By looking at such a long list of potential errors, an organization can effectively identify problematic areas, processes, procedures, or even clinicians and begin addressing them.

Triggers for Electronic Health Record (EHR)-Related Errors

In November 2011 the Institute of Medicine released a report titled "Health IT and Patient Safety: Building Safer Systems for Better Care" in which they highlighted the possibility that the health information technology (HIT) (e.g., electronic health records or barcode medication administration systems) itself could be a source of potential patient harm. This new type of "HIT-related error occurs anytime the HIT system is unavailable for use, malfunctions during use, is used by someone incorrectly, or when HIT interacts with another system component incorrectly, resulting in data being lost or incorrectly entered, displayed, or transmitted."[12] While HIT is often seen as a solution to potential patient safety issues, clearly, we need methods of measuring and monitoring the systems themselves to ensure that they are safe for use.

Toward that end, in 2008, Koppel et al. developed an automated trigger to identify potential medication ordering errors.[13] Briefly, their trigger looked for all medication orders that were canceled by the ordering physician within 2 hours of their entry. They hypothesized that these "rapid discontinuations" represented orders that were subsequently determined to be suboptimal, or in error, in some way. To evaluate their system they reviewed all orders meeting this criteria over a 24-day period (total = 398). Upon review, two-thirds of the orders discontinued within 45 minutes were deemed inappropriate in some way (e.g., often these orders were entered on the wrong patient).

In another example of an automated trigger used to identify errors related to use of the EHR itself, Wilcox et al. were able to estimate the prevalence (or rate) of wrong patient notes (i.e., clinical notes judged to pertain to a different patient than the one they were stored under) in the electronic medical record.[14] Using their EHR, they identified cases in which the demographic information (i.e., patient gender) contained in the free-text portion of the progress note, differed from that contained in the coded portion of the patient's record. They did not find a large percentage of mismatched patient records (e.g., approximately 0.5%), but the trigger is still important. They also realize this estimate represents only half of the likely errors, since wrong patient errors are just as likely in patient notes in which the gender matches.

Conclusion

By combining the clinical and administrative information contained in state-of-the-art electronic health records with the newfound clinical analytics capabilities of modern data warehouses, clinical informatics researchers are poised to revolutionize the study of patient safety. These new computing systems with their automated algorithms are capable of sifting through thousands of patient records to identify potential clinical errors and systematically measure patient safety in ways never before anticipated. Once these measurement systems are in place, informatics researchers can better evaluate the effect of the myriad clinical decision support interventions currently under development. Only after all these new capabilities are in place and functioning as anticipated, can we expect to see the transformative improvements in patient safety that every patient deserves.

Notes

1. Kilbridge P. M., Classen D. C. The informatics opportunities at the intersection of patient safety and clinical informatics. *J Am Med Inform Assoc.* 2008 Jul-Aug;15(4):397-407.

2. Kohn L., Corrigan J., Donaldson M., eds. To Err Is Human: Building a Safer Health System. Committee on Quality of Healthcare in America, Institute of Medicine. National Academies Press; Washington, DC; 1999.

3. Jha A. K., Classen D. C. Getting moving on patient safety— harnessing electronic data for safer care. *N Engl J Med.* 2011 Nov 10;365(19):1756-8.

4. Kahn M. G., Ranade D. The impact of electronic medical records data sources on an adverse drug event quality measure. *J Am Med Inform Assoc*. 2010 Mar-Apr;17(2):185-91.

5. Newman-Toker D. E., Pronovost P. J. Diagnostic Errors—The Next Frontier for Patient Safety. *JAMA*. 2009;301:1060–1062.

6. Chandra A., Nundy S., Seabury S. A. The growth of physician medical malpractice payments: evidence from the National Practitioner Data Bank. Health Aff (Millwood) 2005;(Suppl Web Exclusives):W5-240–W5-249.

7. Singh H., Thomas E. J., Khan M. M., Petersen L. A. Identifying diagnostic errors in primary care using an electronic screening algorithm. Arch Intern Med. 2007 Feb 12;167(3):302-8.

8. Nuñez S., Hexdall A., Aguirre-Jaime A.. Unscheduled returns to the emergency department: an outcome of medical errors? Qual Saf Healthcare. 2006 Apr;15(2):102-8.

9. Classen D. C., Resar R., Griffin F., Federico F., Frankel T., Kimmel N., Whittington J. C., Frankel A., Seger A., James B. C. "Global trigger tool" shows that adverse events in hospitals may be ten times greater than previously measured. Health Aff (Millwood). 2011 Apr;30(4):581-9. Erratum in: Health Aff (Millwood). 2011 Jun;30(6):1217.

10. Classen D. C, Pestotnik S. L., Evans R. S., Burke J. P. Computerized surveillance of adverse drug events in hospital patients. *JAMA*. 1991 Nov 27;266(20):2847-51.

11. Szekendi M. K., Sullivan C., Bobb A., et al. Active surveillance using electronic triggers to detect adverse events in hospitalized patients. Qual Saf Healthcare 2006. 15184–190.

12. Sittig D. F., Singh H. Defining health information technology-related errors: new developments since to err is human. *Arch Intern Med*. 2011 Jul 25;171(14):1281-4.

13. Koppel R., Leonard C. E., Localio A. R., Cohen A., Auten R., Strom B. L. Identifying and quantifying medication errors: evaluation of rapidly discontinued medication orders submitted to a computerized physician order entry system. *J Am Med Inform Assoc*. 2008 Jul-Aug;15(4):461-5.

14. Wilcox A. B., Chen Y. H., Hripcsak G. Minimizing electronic health record patient-note mismatches. *J Am Med Inform Assoc*. 2011 Jul-Aug;18(4):511-4.

15

Using Advanced Analytics to Take Action for Health Plan Members' Health

Stephan Kudyba, Thad Perry, and John Azzolini

"Hot spotting" has practically become the battle cry for managed care organizations since an article of the same name was published in *The New Yorker* in January 2011 (Gawande, 2011). In this article a physician, Jeffrey Brenner, realized after attempting to save a shooting victim it was possible to study healthcare utilization patterns in much the same way assault and crime patterns are tracked. For example, Brenner discovered that across a six-year period, more than 900 people in Camden, New Jersey, accounted for more than $200 million dollars in healthcare expenses. One of these patients had 324 admissions in five years while another patient cost $3.5 million. Additionally, he found that 1% of the patients who used Camden's healthcare facilities accounted for 30% of the total costs.

Called the "Top 1%" by the managed care industry, Brenner's observation was important not because it was new information (e.g., Stanton & Rutherford, 2005), but because it clearly articulated this situation to the general public. Patients like these, "hot spots," were the ones in need of better quality of care. From his experience as a physician, he knew that those patients with the highest costs are usually the ones who experience the worst care. Though hardly a new idea, in less than a year since this article was published, the concept of identifying and intervening with the most severe or clinically complex patients to lower future healthcare costs has seen resurgence in the managed care arena.

State Medicaid agencies have embraced the hot spotting concept, challenging their Medicaid Managed Care Plans (MCPs) to decrease overall healthcare expenditures by increasing the quality of care for their member populations, especially those who are currently a significant cost burden to the program. As early as 2005, the Ohio Commission to Reform Medicaid (OCRM) published a report on Ohio Medicaid making recommendations and outlining action steps focused on (a) long-term care, (b) care management, (c) pharmacy management, (d) member eligibility, (e) financial administration, and (f) overall program structure and management (Revisiting Medicaid Reform, 2009). Recognizing the difficulty of healthcare reform, this report, and subsequent study, underscored the complexity of the relationships between quality, delivery, management, administration, cost, and payment of healthcare.

Clearly, quality of care and cost of care are on a collision course. Is it possible to develop a system that establishes and supports our nation's need for quality, affordable healthcare, or has decades of fee-for-service payment systems independent of quality of care and/or patient outcomes made it impossible to reform heath care delivery? Unfortunately, as long as healthcare facilities, providers, and suppliers fail to work together and independently treat patients as profit centers, there appears to be little hope for healthcare reform.

This is one reason that the new healthcare reform law is so controversial. It is widely known that health plan members who receive the right type of care at the appropriate time have better health outcomes, thereby decreasing their cost burden on an already overstressed healthcare economic system. Proponents and critics of healthcare reform do agree on one thing—there are ways to decrease overall healthcare costs through population-based care management programs. Brenner asserted in *The New Yorker* article, "For all the stupid, expensive, predictive-modeling software that the big vendors sell, you just ask the doctors, 'Who are your most difficult patients?,' and they can identify them." Though this statement has simplistic appeal, it cannot exist as a management strategy applicable to broad populations. The process of designing and running a population-based program is not as trivial as this statement implies.

To be effective, a data-driven, procedural approach must be taken, which allows for the coordination and collaboration of all constituents in the healthcare delivery process. Moreover, robust data management, data mining, and analytic processes are essential for successful population-based programs because improvements in quality and cost of care can only occur when *the right members receive the right services at the right time* (Cousins et al., 2002; Perry et al., 2004, 2007). For these data-driven methods to influence clinical and financial outcomes of our healthcare delivery system, the information derived from them must be *actionable*.

Clearly, care management programs already exist to address these issues. However, one must consider the following questions: If there are already programs, protocols, and processes in place, why are we not seeing significant improvements in quality and cost of care? Why are health plans, facilities, and provider practices still challenged with improving quality of care while reducing overall costs? Therefore, we are not presenting new concepts in care management, but rather a conceptual framework comprised of factors influencing one's overall healthcare experience. This combination of factors allows for the optimization of care management interventions that maximize the ability to target and identify those members whose cost burden will continue to increase in the absence of care management support activities.

To develop actionable care management information, it is necessary to create a framework that takes into account data derived from numerous sources. Expert systems created to support clinical decision support systems (CDSS), are relatively common in facilities and provider offices. Many of these systems are designed to improve (a) patient safety, (b) quality of care, and (c) efficiency of delivery (Coiera, 2003; Raghupathi, 2007). These expert systems are extremely helpful in clinical settings because they provide *actionable information*. That is, the information derived from these systems is directly applicable to improving the treatment of specific patients. If expert systems are successfully implemented at a provider/facility level, then it is reasonable to expect that expert systems can also be implemented at a health plan/managed care level.

Actionable Information—A Conceptual Framework

We propose developing a care management expert system that combines information from three healthcare cost and utilization domains:

- Current and predicted healthcare costs
- Utilization impact
- Member engagement

Using current and predicted healthcare costs, it is possible to identify those health plan members who are currently highest cost and will remain high cost in the absence of healthcare interventions. Since it is known that current high cost does not necessarily predict future high cost (see Ridinger & Rice, 2000), predictive modeling is necessary to determine those individuals with the highest probability of either remaining or becoming high cost health plan members. Without these analytic methods, it would not be possible to direct care management interventions to the right risk groups to significantly influence the overall cost burden of these health plan members. To simplify the challenge, it is helpful to visualize the relationship between current and future healthcare costs using a resource allocation matrix (e.g., Donaldson et al., 2002).

As shown in Figure 15.1, the relationship between current and future cost burden is an important factor in determining where to allocate resources. Simple predictions (i.e., those that do not require involved predictive modeling techniques) allow for the classification of members into either "good use" or "poor use." Likewise, complex predictions (i.e., those that require involved predictive modeling techniques) do the same. It is, obviously, more difficult to determine those low cost members who will become high cost as well as those high cost members who will become low cost. By identifying those future high cost members, the correct resources can be allocated to the appropriate interventions needed to help control healthcare costs. Since the challenge of this care management strategy is to control future healthcare costs, the first step in creating actionable information is to classify the member population.

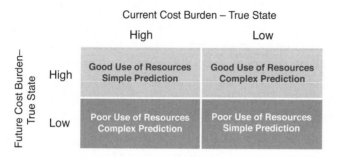

Resource Allocation Matrix

Figure 15.1 Resource allocation matrix

Knowledge Discovery through Multivariate Analytics

Simple database queries can be utilized to search vast data sources to identify current high cost patients in a patient population (e.g., those incurring extensive health treatments, as mentioned in the *New Yorker* article). This information can be beneficial since resources can be applied to those particular cases to address those factors contributing to continued poor health maintenance. One limitation of this approach, however, is that it requires individuals to become high service utilizers or high cost before resources can be applied to alleviate or mitigate the problem. In other words, the limitation of this more hind-sight or reactive analytic method is that the focus is placed on what has already happened, where little knowledge is generated as to why individuals become high cost.

More effective cost reduction and health enhancement policies can be achieved through proactive analytics that identify likely future "poor health" or high cost candidates (Kudyba, 2005). Advanced analytical methods applied to robust data resources enable decision makers to identify individuals at risk of requiring extensive health treatments or high cost candidates. Methodologies such as logistic regression help identify noteworthy patterns existing in corresponding data. These patterns can include variables involving patient demographic and behavioral information, symptomatic and diagnostic data, and treatment related data to name a few. The resulting models identify not only the segments of a patient population that are likely high cost candidates but also provide the possible factors that lead patients to be high cost...or the "why" behind the high cost results.

Advanced analytic methods that incorporate a multivariate approach and the utilization of mathematical and algorithmic processing of data in conjunction with statistical testing techniques are often referred to as "knowledge discovery techniques." The "knowledge" refers to the identification of patterns or relationships between variables in a particular data set that explain "why" things happen, not simply "what" has happened (see Figure 15.2). Perhaps the most important information to extract from data with regards to the complexity of individuals' health status includes patterns in dietary behavior, physical attributes, treatment and medication practices, and so on that can provide insights as to what combination of behavioral and descriptive variables lead to a less healthy, higher cost individual. This information is truly actionable as it empowers healthcare providers to more accurately apply available resources to mitigate costs and maintain a healthier population by implementing preemptive or proactive treatment to high risk candidates, thus mitigating costs before they incur. The result is not only reduced cost for providers but a healthier population.

The next step in the conceptual model involves determining the opportunity to impact a member's healthcare utilization level. This step is multifaceted and requires the integration of both a member's health status as well as their use of healthcare services. Health status refers to the current standing of an individual's clinical, physical, and mental health (Kudyba et al., 2008). Health status is commonly determined through general health assessments (GHAs), health risk assessments (HRAs), clinical severity estimators (e.g., ACGs—adjusted clinical groups), as well as other sources of member-specific healthcare information (e.g., provider records, treatment notes, etc.). This type of information allows care managers to better understand if they can impact these members' utilization behaviors, given their current health statuses. If a member suffers from a disease or condition that cannot be impacted by care management interventions, then decisions must be made as to the level and type of services and support they receive.

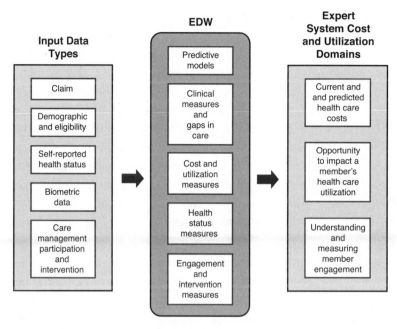

Figure 15.2 Conceptual model for predicting patient costs

Similarly, a member's utilization history must be analyzed to determine the opportunity to modify his behavior. For example, it has been well documented that care management interventions can influence health plan members' emergency department and hospital admission rates (Brandon, 2011). With the current market focus on emergency department diversion (EDD), readmission rates, and prescription drug rates, it is becoming increasingly important to integrate these sources of information. The development of enterprise data warehouses (EDW) at health plans is a direct result of the need to collect and use these data sources to track their members' utilization behaviors. As described earlier from the "Hot Spots" article, when a health plan member has 324 admissions in a five-year period, one has to ask if this level of utilization was necessary. Consequently, the second step of this process involves an assessment of health status and utilization history, which further filters the membership population by taking those members who are predicted to be future high cost and identifying those who can be impacted by care management interventions.

Understanding and measuring member engagement is the final step in this conceptual model. Engagement can mean many different things, but in this context it refers to a member's willingness to transmit, receive, and act upon customized communications. Engagement is emotional in nature; it often describes one's connection with healthcare providers and predicts how actively members participate in the management of their own healthcare (Arnold, 2007). In other words, engagement involves members' management of their own healthcare services to meet their healthcare needs. When members are engaged, they are actively involved and focused on their healthcare behaviors, resulting in better choices, better provider communications, and, ultimately, better outcomes (Protheroe, 2008).

Patient adherence to treatment regimens is a good example of why engagement is important in this conceptual model. For example, glycemic control is essential to individuals suffering from either Type I or Type II diabetes. This condition requires a high level of self-management; in other words, highly engaged individuals with diabetes have much better outcomes than those who are not engaged. An engaged individual will most likely adhere to the treatments prescribed by his providers (Delamater, 2006; Milano, 2011). Therefore, when members are defined by their future high cost, impact of care management interventions, and engagement in healthcare, the resulting actionable information increases the opportunity to provide *the right members, the right services, at the right time*.

Conclusions

There is no mistaking that population-based care management programs are difficult to develop, implement, and manage. A few of the more pervasive issues are as follows:

- Healthcare data sources are seldom standardized, often full of errors, and difficult to aggregate.
- Privacy and security regulations add additional complexity to data management efforts.
- Administrative claims data are collected on multiple systems and have varying file formats and record layouts.

- Health plan membership constantly changes, with members entering and exiting the program based on employment and eligibility status.

- Provider enrollment is extremely variable based on participation status, physical location, and professional specialty.

- The prevalence of chronic conditions continues to rise at an alarming rate.

- Treatment and drug regimens continually evolve as well as practice guidelines, evidence-based medicine, and medical technology.

- Healthcare reform brings new concepts such as health homes, accountable care organizations, and health insurance exchanges.

The preceding list only scratches the surface. However, as one contemplates current healthcare delivery challenges, a common theme surfaces. All of these issues have a singular dependency— *actionable information.*

Figure 15.3 represents an elementary formulation of the conceptual framework proposed in this chapter. The input to this model (multiple data sources) continues to be an important topic by itself. Data management, data aggregation, and data warehousing of the multitude of disparate healthcare data sources must occur before any other activity can take place. The scope and magnitude of this effort cannot be overemphasized. In order to effectively manage health plan members at a population level, administrative claims data sources (e.g., medical claims, pharmacy claims, durable medical equipment and supply claims, laboratory claims, behavioral health claims, dental claims, home health claims, skilled nursing and nursing home claims, etc.), health and risk assessment data sources (e.g., general health assessments, health risk assessments, disease specific surveys, etc.), risk and grouping data sources (e.g., episode treatment groups, diagnosis related groups, adjusted clinical groups, etc.), as well as many other healthcare data sources must be combined in an understandable, functional way. Health plans and business intelligence vendors are working on this challenge, with varying degrees of success as they work towards building enterprise data warehouses (EDW). This is the healthcare industry's biggest challenge because without clean, robust data, it is not possible to implement data-driven care management programs, not to mention all of the new processes and programs created by the Affordable Care Act (ACA).

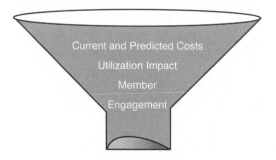

Figure 15.3 A simple formulation of the conceptual framework

Once data are entered into the system, the process of identifying the best candidates for care management programs begins. Following the proposed framework, each of the three categories acts as a population filter. Once the target population is identified (e.g., members with asthma), current and predicted cost burden is calculated for each individual member. The result of this analysis is a ranking of all members with asthma (continuing with this example) from highest to lowest predicted cost. The next step involves assessing the estimated utilization impact of the care management program's interventions on these members with asthma. Those members who are determined to be most "impacted" by the care management program are ranked higher than those who are not. These two sources of information (predicted cost burden and utilization impact) are then combined to produce a new risk ranking; this ranking takes into account those members with the highest predicted cost burden who also have the highest probability of utilization impact by the care management program.

Combining these two sources of information gives care managers not only the ability to identify those members with the highest predicted costs but also those members they will have the best opportunity to impact. Consequently, a member with high predicted cost with low utilization impact would be ranked lower than a member with a moderately high predicted cost with a high utilization impact. Using this methodology, the entire population of members with asthma can be ranked from highest to lowest risk based on predicted costs and utilization impact. As described previously, this actionable information allows care managers to make appropriate resource allocation decisions. Since it would be a poor use of resources to concentrate efforts on high cost members with low to no probability of utilization impact, this information would be valuable for program management.

Nevertheless, this proposed framework has one final filter—member engagement. Once the member population is identified and risk ranked by predicted cost and utilization impact, member engagement is assessed. It does not matter how high one's financial risk might be or how impacted she could be by care management interventions if the member is not engaged in the process. If a member is not emotionally invested in the program, there is little chance that they will benefit from the care management interventions. Without engagement, care management does not work. Consequently, at a population level, using engagement as the last filter will further refine the information derived from the model, thereby increasing the chances that care managers are expending efforts on those members who will be high future cost, have high utilization impact opportunities, and are engaged in the care management process. This framework, comprised of three information filters, will identify those members who, if left unattended, will present the highest cost and utilization risk and are able to engage successfully in the program. This information is essential at a population management level because limited care management resources are available and correct resource allocation is paramount to program success.

Clearly, population management programs rely on data-driven methods to target populations, identify high risk members, allocate limited resources, and assess program outcomes. To increase quality of care while decreasing healthcare costs, care management programs must focus their efforts on intervening with the right members, with the right services, at the right time. Once data are transformed into actionable information, program successes are possible, resulting in positive care management utilization and financial outcomes. Possessing data is not enough—actionable information must be created and used by health plans to meet current and future challenges. The conceptual model proposed in this chapter is one example of how the thoughtful consolidation of information can assist in the correct allocation of resources in a population-based care management program.

To be effective, a data-driven, procedural approach must be taken, which allows for the coordination and collaboration of all constituents in the healthcare delivery process.

References

Arnold, S. (2007). Improving Quality Healthcare: The Role of Consumer Engagement. *Robert Wood Johnson Foundation Issue Brief*. October 2007.

Brandon, W. (2011). Reducing emergency department visits among high-using patients. *Journal of Family Practice*. FindArticles.com. 15 Nov, 2011.

Coiera, E. (2003) *The Guide to Health Informatics*. 2nd Edition: Arnold, London.

Cousins, M., Shickle, L., and Bander, J. (2002). An Introduction to Predictive Modeling for Disease Management Risk Stratification. *Disease Management* 5: 157-167.

Delamater, A. (2006). Improving Patient Adherence. *Clinical Diabetes*. 24(2): 71-77

Donaldson, C., Currier, G., and Mitton, C. (2002). Cost Effectiveness Analysis in Healthcare: Contraindications. *British Medical Journal*. BMJ; 325:891.

Gawande, A. (January 24, 2011). The Hot Spotters: Can We Lower Medical Costs by Giving the Neediest Patients Better Care? *The New Yorker*.

Kudyba, S., Perry, T., and Rice, J. (2008). Informatics Application Challenges for Managed Care Organizations: The Three Faces of Population Segmentation and a Proposed Classification System. *Int. J. of Healthcare Information Systems and Informatics*. 3(2): 21-31.

Kudyba, S., Hamar, B., and Gandy, W. Enhancing Efficiency in the Healthcare Industry, *Communications of the ACM*, December 2005.

Milano, C. (2011). Can Self-Management Programs Ease Chronic Conditions? *Managed Care* January 2011. MediMedia USA.

Perry, T., Tucker, T. et al. (2004). The Application of Data Mining Techniques in Health Plan Population Management. Chapter VII. *IT Solutions Series: Managing Data Mining*. Idea Group Publishing.

Perry, T., Kudyba, S., and Lawrence, K. (2007). Identification and Prediction of Chronic Conditions for Health Plan Members using Data Mining Techniques. In *Data Mining Methods and Applications*. New York: Taylor & Francis: 175-182.

Protheroe, J., Roger, A., Kennedy, A., MacDonald, W., and Lee, Victoria (2008). Promoting Patient Engagement with Self-Management Support Information: A Qualitative Meta-Synthesis of Processes Influencing Uptake. *Implementation Science*. 3:44.

Revisiting Medicaid Reform: A project of The Center for Community Solutions and the Center for Health Outcomes, Policy, and Evaluation Studies (HOPES) of Ohio State University. (2009). A Status Report on Recommendations from the Ohio Commission to Reform Medicaid Four Years Later: January 2005-January 2009.

Raghupathi, W. (2007). Designing Clinical Decision Support Systems in Healthcare: A Systematic Approach. *International Journal of Healthcare Information Systems and Informatics*. 2(1): 44-53.

Ridinger, M. and Rice, J. (2000). Predictive Modeling Points Way to Future Risk Status. *Health Management Technology*. 21: 10-12.

Stanton, M., and Rutherford, M. (2005). The High Concentration of U.S. Healthcare Expenditures. Agency for Healthcare Research and Quality. *Research in Action*, Issue 19. AHRQ Pub. No. 06-0060.

16

Measuring the Impact of Social Media in Healthcare

David Wiggin

The intersection of social media and the U.S. healthcare system is rapidly expanding. The goal of this chapter is to present a snapshot of current and emerging uses of social media and take a step toward the development of an analytical model to measure the impact of social media on U.S. healthcare. This becomes most interesting if social media can play a *measurable* role in improving the health of the population.

Why Measure at All?

> When you can measure what you are speaking about, and express it in numbers, you know something about it, when you cannot express it in numbers, your knowledge is of a meager and unsatisfactory kind; it may be the beginning of knowledge, but you have scarcely, in your thoughts advanced to the stage of science. —Lord Kelvin

Lord Kelvin was on to something here. If we can quantify the value of social media, we can make informed decisions about social media investments. Today, social media investments are being made in healthcare to increase market share in competitive markets, improve quality of care, reduce the cost of care, and reduce health risk. Ultimately, the desired outcome of all these investments is to improve the health of the population. While we start by measuring first order impacts like market share and cost, we should seek ways to measure impact on population health.

Working Definition of Social Media in Healthcare

A paper written by the Computer Services Corporation (CSC) in March 2012 offered this definition: "Social media is the process of people using online tools and platforms to share content and information through conversation and communication."[1]

Most of us probably think about Facebook and Twitter first when we hear the phrase social media. However when we consider the impact of social media on healthcare, we need to consider the powerful *combination effect* of these channels with

- **Blogs**—Most often by physicians, sometimes by patient advocates
- **Affinity group sites**—Condition-specific or role-specific
- **Reference sites**—Dot-coms, crowd sourced sites, patient experience rating sites

While at first blush, reference sites may seem to not belong in this conversation, they are in fact a vital part of the social media landscape. So much of what happens in social media includes a link to content from another site. Twitter is a hundred times more powerful and interesting because tweets are link-enabled to content. The key point from this definition is that social media analytics must take into account the constellation of connected content.

The Complexity of Social Media and Healthcare

Before we launch into a conversation about *how* social media is being used in healthcare today, let's first consider the relative complexity of the topic.

If social media is the connection point of two or more parties, then the number and complexity of relationships is important to understand. Let's first consider the intersection of social media and consumer packaged goods (CPG). There we have businesses (Proctor & Gamble, Unilever, Colgate-Palmolive), and we have consumers (you and me). In this case, the product-to-consumer relationship may be well understood. There are many channels connecting CPG companies to consumers, each with a measurable impact—distribution channels, advertising channels, and so on.

Healthcare is more complicated because more than two primary parties are involved, thereby generating more relationships, each in turn subject to connection via a variety of channels. If that weren't enough, there is a lot of variety among healthcare patients/consumers, including varying conditions and goals, the complications of multiple conditions, many available provider options, regional variations in care delivery, and uncertainty in outcomes.

Each of these factors increases the complexity of the relationships. Let's take a look at these complexity multipliers. They include

- Five major roles—patients, physicians (and other outpatient care), hospitals, payers (employers, health plans), and health IT
- 500 categories of chronic and acute conditions (episodes of care) with more granular detail available via 68,000 diagnosis codes (ICD-10 CM)
- Variations in the practice of medicine regionally, and by individual physicians (700,000) and hospitals (5,700) within a region
- Variations in patient behavior based on socioeconomic, cultural, and regional influences (compliance with treatment, lifestyle choices, and avoidable chronic disease)

It's one thing to measure the social media impact of notable Facebook campaigns like those of American Express, Red Bull, Lacoste, and Petco. However, when it comes to the significantly more complicated industry landscape in healthcare, we should expect the analytics at the intersection of social media and healthcare to also be more complex.

Who Is Involved in Each Category of Social Media Use Today?

As we continue to understand the complexity of healthcare social media (HCSM) analytics, Figure 16.1 takes a step toward segmenting that complexity among interactions. It summarizes the number of roles (or nodes) in the social network by purpose of the connection.

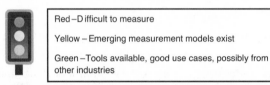

Red – Difficult to measure

Yellow – Emerging measurement models exist

Green – Tools available, good use cases, possibly from other industries

	Purpose	Patient	Physician	Hospital	Health IT	Payer	Pharma Companies
1	Marketing, Advertising	✔	✔	✔	✔	✔	✔
2	Patient recruiting (clinical trials)	✔	✔				✔
3	Patient Reviews / Provider Reputation	✔	✔	✔		✔	
4	Provider Collaboration / Education		✔	✔	✔		
5	Patient Education	✔	✔		✔	✔	✔
6	Patient Affinity Groups	✔			✔		✔
7	Patient Monitoring	✔	✔	✔	✔		
8	Care Management	✔	✔	✔	✔	✔	

Figure 16.1 Nodes in the social network by connection for HCSM

The colors categorize our readiness to measure the social media interactions described by the interaction. The check marks identify which role (stakeholder) is involved in that connection.

What Analytics Are Enabled Today and in the Future?

In Figure 16.1, the first three categories of social media use and the analysis of these areas have the benefit of being comparable to work that has already been done in other industries. Marketing effectiveness, research, sentiment analysis, and brand management are all well-established disciplines. For example, in an article titled "Brands Ignore Negative Social Buzz at Their Peril," eMarketer.com, reported "that 46% of US internet users it surveyed had turned to companies' social media sites to vent their frustrations about poor experiences."[2] One healthcare industry analyst commented that the healthcare industry would do well to learn from the retail and banking industries when it comes to the use and analytics of social media.[3] For the purposes of this chapter, we focus on social media uses that are unique to healthcare.

The fourth category in Figure 16.1, Provider Collaboration/ Education, carries big potential in encouraging innovation and change in the physician community. In this use of social media, a growing collection of healthcare providers have formed a broad network of specialists, whose most advanced practitioners are sharing their findings and teaching others.

Social media provides clinicians with a convenient way to follow the work of specialists, researchers, and other peers, with an investment of just a few minutes a day. Using Twitter, they follow individuals (@David_Wiggin) and topics (#HCSM), and receive the benefit of articles, blog posts, and other materials that are germane to their professional lives. This will supplement, if not replace, having to wade through journals and online magazines and maintaining multiple e-mail subscriptions and feeds. They can follow the work of key opinion leaders and others whose views they trust. The result is a social network of clinicians who are continuously learning and sharing, and who have entered into a form of informal professional accountability. While much of this information is available via traditional sources, many find the nature of the interaction to be more manageable and helpful due to the shorter time between question and answer.

To give you an idea about how vigorous this community is, there's a Web site called twitterdoctors.net. Every hour, it provides an updated list of the most influential physicians on Twitter. The number of followers ranges from thousands to millions, and the number of tweets is in the thousands to tens of thousands.

Measuring the impact of this use of social media will be tricky. Measuring influence, as this Web site does, is interesting, but it doesn't explain how influence correlates to the delivery of high quality, coordinated care, or how costs have been reduced based on participation. So while counting tweets is easy and has some relevance, the more meaningful measures would be longer term improvements in the quality of care delivered and patient outcomes. Any analytics that demonstrates the improvement in the health of the population via the use of social media would be valuable.

The last four categories shown previously in Figure 16.1 are all focused on patient health. Each can play a role in transforming the health crisis.

The fifth category, Patient Education, is interesting because social media is another channel for distribution of relevant healthcare content, usually by providers. Articles, video, and other content are pushed via a social media channel. For example, if a provider can send a YouTube video link to her patients, it has the potential to encourage appropriate patient lifestyle choices and leverage the physician-patient interaction at the office visit. In a sense, this is the social media upgrade to the printed brochure that you used to take home from the doctor's office. In addition to the office visit conversation, physicians are directing patients to this education content via e-mail, Facebook messages, and by blogging.

The purpose of social media in this context is targeted communication with the goal of changing patient behavior. The following summary of population health analytics acknowledges the importance of Web-based and mobile solutions, which is a category aligned with social media:

> Population health is holistic in that it seeks to reveal patterns and connections within and between multiple systems and to develop approaches that respond to the needs of populations. It defines wellness as more than the absence of illness and seeks to improve physical, emotional, and behavioral fitness. Population health tactics include rigorous analysis of outcomes. Understanding population-based patterns are critical antecedents to addressing population needs. That is, data informs the selection of effective population health

management strategies to prevent or diminish illness in the future. Increasingly these efforts include the integration of face-to-face clinical care, telephonic support through coaching and care management, as well as web-based and mobile solutions, which support greater health literacy and consumerism."[4]

The best way to measure the effectiveness of this use of social media is through the use of patient surveys. If the goal is to measure the beneficial impact of patient education, it will be important to hear from the patients. Only a percentage of patients who received brochures read them, and even a smaller percentage followed the guidance in the brochure, yielding a health benefit. The same is true with social media. Just because a link was sent or a video was watched doesn't mean that education and lifestyle change took place.

The sixth category, Patient Affinity Groups, forms a leveraged crowd-sourcing opportunity for consumers to make informed choices about treatment and to learn from others who have been diagnosed with a condition. Further, when they achieve critical mass (enough participants) they collect self-reported patient data about what works and what doesn't to form a growing self-reported comparative effectiveness study. This is in fact the business model for several Web companies today who manage these sites.

www.patientslikeme.com is a great example of a social network tool in this category. Here's their story:

> PatientsLikeMe was co-founded in 2004 by three MIT engineers: brothers Benjamin and James Heywood and longtime friend Jeff Cole. Five years earlier, their brother and friend Stephen Heywood was diagnosed with ALS (Lou Gehrig's disease) at the age of 29. The Heywood family soon began searching the world over for ideas that would extend and improve Stephen's life. Inspired by Stephen's experiences, the co-founders and team conceptualized and built a health data-sharing platform that they believe can transform the way patients manage their own conditions, change the way industry conducts research and improve patient care.

So far, more than 150,000 people have subscribed and are able to learn about the experiences of others in managing symptoms, understanding the progression of disease, observing the effectiveness of various treatments, and finding moral support and encouragement from others who suffer from the same condition.

These sites also match patients with clinical trials, supporting both research and healing.

This is another case where it would be complicated to quantify and appropriately attribute the value of participating in this site across the whole participant pool. There will be many anecdotal stories that will continue to encourage participation, but isolating the benefit of this site to a population health measure will be difficult.

The seventh category, Patient Monitoring, takes advantage of the latest inventions and IT gizmos to help each of us stay on track in our lifestyle choices. The relatively new quantified-self movement grew out of the idea that if we could more easily track our behavior and even (optionally) join others in their quest to improved health, improved health would result.

Two great examples of this technology are the fitbit and Nike's fuelband. Each records biometric information and transmits to your smartphone, providing you with data about exercise, calories burned, sleep, and so on. When combined with self-reported data like calorie consumption, individuals make a more informed choice about whether to snack at 2:00 p.m. and/or whether to have dessert that evening.

The potential for analytics on all this sensor data is great, and on an individual basis, this is included in the free smartphone app when you buy the $99 fitbit or similar device. Beyond the individual data, in the not too distant future, could we envision sharing some of these data with our accountable care organization (ACO) or health plan in exchange for a lower cost plan? From a patient/member perspective, this could be like the discount we receive for voluntarily filling out the Health Risk Assessment or placing a GPS equipped device in our automobiles to earn a lower safe driver rate from our auto insurance company.

This category of use makes the social media category because of the significant social dimension to the use of these technologies.

Users are encouraged by the manufacturers and their peers to share their progress and compete with others, both of which fuel continued use and healthy lifestyle changes. This allows users to sign up for peer pressure and accountability that has a fun dimension.

The eighth category, Care Management, and its more focused predecessor, disease management, has long been recognized for its value in providing extra support to patients to ensure that they receive the care they need. Social media has made it possible for payer and provider organizations to interact with their members/patients on more channels, with more tailored communications.

Care management started with phone calls and letters in the mail. With broad adoption of home computers and cell phones, these were replaced with e-mail reminders and text messages. The Association of Managed Care just published a study touting the effectiveness of a text message program to encourage patient compliance in medications and follow-up visits.[5] Some providers have started to contact their patients via Facebook. Last summer, the University of Iowa Children's Hospital launched a program to improve medication adherence on Facebook.[6]

Analytics on the cost/benefit of care management programs are mature and well understood today. The use of social media channels is a natural extension of earlier programs, and should plug into the existing analytics framework easily.

Conclusion

The use of social media in the context of healthcare is growing and shows no signs of slowing down. It has been the topic of many conferences this year (Duke, NY Medical College, Mayo Clinic, etc.), and it is a promising collection of technologies that form a complicated, yet effective, channel that is accessed by all stakeholders.

However, the ability to attribute benefits to specific uses of social media varies from situation to situation. While we will always seek to do that, it's important to realize that the value of social media may only be detected via patient survey (usage survey, satisfaction survey) or more importantly in measurements of the improvement of population health.

The best practice approach to analytics of this type continues to be collect detail data from relevant sources as it is generated, integrate the data for a comprehensive view of the patient, and put the data to use immediately via the appropriate channel(s)—social media in this case.

While it will be difficult to measure the long-term value of a social media investment with any precision, the short-term business value of these investments should be measured using traditional business metrics like market share (new customers, customer retention), sentiment analysis, and customer satisfaction.

Notes

1. "Should Healthcare Organizations Use Social Media? A Global Update," CSC, March 2012, p. 4. http://assets1.csc.com/health_services/downloads/CSC_Should_Healthcare_Organizations_Use_Social_Media_A_Global_Update.pdf.

2. "Brands Ignore Negative Social Buzz at Their Peril," eMarketer.com, July 16, 2012.

3. Janice Young, "The Healthcare Consumer We Hardly Know," IDC Health Insights, April 4, 2012.

4. Raymond Fabius, MD, CPE, FACPE; Linda MacCracken, MBA; Jill Pritts, MS, "Vocabulary of Healthcare Reform," January 2012, p. 39. Thomas Reuters. www.ahip.org/WP-VocabHCReform/.

5. Henry H. Fischer, MD; Susan L. Moore, MSPH; David Ginosar, MD; Arthur J. Davidson, MD, MSPH; Cecilia M. Rice-Peterson, RN, BSN; Michael J. Durfee, MSPH; Thomas D. MacKenzie, MD, MSPH; Raymond O. Estacio, MD; and Andrew W. Steele, MD, MPH, MSc, "Care by Cell Phone: Text Messaging for Chronic Disease Management," *Am J Manag Care*, 2012;18(2):e42-e47.

6. Ken Terry, "Many Doctors Don't Take Social Media Beyond Marketing," Informationweek.com, April 10, 2012. http://www.informationweek.com/healthcare/patient/many-doctors-dont-take-social-media-beyo/232900043.

Part IV
Best Practices in Healthcare Analytics Across the Ecosystem

17

Overview of Healthcare Analytics Best Practices Across the Ecosystem

Dwight McNeill

Analytics in healthcare is old and new. Science has been a strong underpinning of healthcare in the research devoted to the discovery of causes and treatments of disease. However, delivering this knowledge from the bench to the bedside to optimize the care of every patient has been an ongoing challenge. Although treating sickness, that is, the interaction between a patient and her caregivers, is the raison d'être of healthcare, the industry is more complex than that. The American way of healthcare requires large doses of payment, finance, regulation, research and development, and administrative and business supports. Healthcare is a huge part of the U.S. economy, accounting for 18% of GDP at a spending rate of $8,500 for every man, woman, and child.[1] And it is big business. Annual hospital expenses are approaching $1 trillion, and physician services are more than $0.5 trillion. Both of these categories of providers amount to more than 50% of spending. The next highest spending area is for prescription drugs, which amounts to 10% of spending.

Healthcare is both an informational and a personal business. It is personal because it deals with people, and communications and relationship skills are fundamental to making change happen. It is informational in that it is about discovery, measurement, improvement, and running a business.

Analytics is the high octane fuel to feed the thirsty information engines. It holds the promise to improve people's lives, increase revenues and reduce costs, and to change the very nature of what healthcare is and what it can be.

Part IV is on best practices and includes eight case studies of leading organizations in healthcare analytics. These are bellwether organizations and represent the best of the art and science of analytics as of 2012. The case studies are inclusive of the settings where analytics is practiced including providers, payers, and a life sciences company. It includes both the public and private sectors.

The chapters in part IV address the "whats" and "hows" of analytics to support organizational strategies and goals. The whats include the domains of the content of analytics, including clinical, business, and marketing purposes. The hows include the functions of analytics, including how it is organized, how it adds value, and its technical challenges.

Providers

Providers are the boots on the ground in healthcare: the doctors, nurses, and a myriad of other care professionals who interact with patients and their families to treat them. Their efforts are important to people's well-being, and very often spell the difference between life and death. This is where the rubber meets the road, where the clinical knowledge created by research, the pills and devices developed by life science companies, the financial coverage provided by insurance companies, the healthcare benefits provided by employers, and the support by all the business functions of hospitals and other healthcare settings all come into play. As such it is a vitally important fulcrum for analytics to support clinicians with information, knowledge, and the tools to improve practice.

The providers included in the case studies are very large, ranging in revenues from just under $1 billion to over $150 billion. They are integrated delivery systems that provide a continuum of care from hospital to outpatient care in an organized and coordinated way and are accountable for the populations they serve both clinically and fiscally. Because they are accountable, they are more incentivized to use analytics and make continual improvements.

The Whats

These five best practice providers in Part IV, including Partners Healthcare System, Catholic Health Initiatives, Veterans Health Administration, Air Force Medical Service, and HealthEast Care System, have been at the vanguard of health analytics partly because of four common areas of content and focus:

- They were early adopters of electronic health records (EHRs). EHRs support their patient care strategies, such as care coordination, disease management, and use of care protocols, by increasing the availability of individual patient and population data and by improving communication among providers. Partners Healthcare, the VA, and the Air Force had EHRs in place systemwide by the early 1990s. Note that only 35% of U.S. hospitals had adopted EHRs by 2011.[2] Early wins from EHRs included Computerized Provider Order Entry (CPOE) systems, which improved the accuracy of physicians. medication orders, and also measuring adherence to medication guidelines. For example, adverse drug events were cut in half after the introduction of these systems at a Partners hospital.

- The leadership for clinical analytics is clear at these organizations. They want to achieve clearly articulated institutional goals such as reducing medical errors, achieving uniformly high clinical quality, improving chronic disease management, and using clinical resources efficiently. One of the keys to the transformation of the VA was a performance measurement system that was used to hold senior managers accountable for improvement in performance measures. The analytics undergirding the accountability system include tracking metrics and reporting on them through dashboards. Similarly, HealthEast set out on a "quality journey" to become the benchmark for quality in the Twin Cities area and deployed analytics for measurement and improvement strategies.

- They use a clinical data warehouse for research purposes. For example, Partners uses it for postmarket surveillance to detect problems with drugs and medical devices after they are released to the market. The VA detected an outbreak of a rare form of pneumonia and was able to determine that a certain nasal spray was the cause. The warehouse also provides the data foundation for supporting many forms of research, which can garner big revenues for these institutions.

- Finally, these institutions use analytics for business and finance functions including optimizing revenue, understanding employer attrition, claims adjudication, and reimbursing physicians based on performance metrics such as cost-effective use of imaging services.

The Hows

Much of the work in analytics in healthcare today is building capacity specifically related to connecting the data "pipes" and integrating the data including various forms of clinical, operational, and financial data. The Partners case study demonstrates the issues involved in deciding what goes into an enterprise level analytics design versus a hospital-specific design. Similarly, the overarching question of Chapter 19, "Catholic Health Initiatives," is how does a healthcare organization translate data into actionable information for every stakeholder across the enterprise? The need for an efficient and scalable data warehouse is discussed in Chapter 21, "The Health Service Data Warehouse Project at the Air Force Medical Service (AFMS)." It addresses the challenges of finding, acquiring, improving, and integrating data and reducing long lead times and frustration on the part of users. Chapter 22, "Developing Enterprise Analytics at HealthEast Care System," focuses on how to organize analytic teams at different levels to accomplish different purposes.

Payers

Payers, including a multitude of commercial health insurers, employers, and governments, provide the financing for the high cost of health services in the United States. Payers face epic challenges, including the advent of health information exchanges, health insurance exchanges, new Affordable Care Act (ACA) regulations on coverage, premium reviews, profit margins and mandates, new provider models such as Accountable Care Organizations (ACOs), and a huge new pool of customers who were previously uninsured. Payers also face the dual demands to 1) change their business model of providing wholesale insurance to employers to providing retail health and insurance services to individuals while also 2) focusing on the health and management of populations.

Payers had gotten into a routine of managing the economics of benefits and coverage, premium pricing, and various insurance products, but were not as actively engaged in managing health and medical care of members/employees as they were for the brief but noteworthy managed care era of the 1990s. Now, the tenor has changed and the pendulum has swung back and beyond such that insurers are changing the very nature of their business by blurring the lines between payers and providers to ensure better business results and also by changing their mission to become a health company and/or to become an information company where insurance is just one product line for these organizations.

The analytics challenges and opportunities are daunting. Payers have relied on claims data as their intelligence source to understand their business but will need to rely on diverse data to address the above challenges. This diverse data coupled with need to comply with unrelenting regulations will necessitate the review of legacy systems, the capacity of existing data warehouses, and a heightened need to integrate, process faster, discover insights, and contribute regularly to the bottom line.

There are two chapters in Part IV on payers, including a health insurer—Chapter 23, "Aetna," by Kyle Cheek—and an employer—Chapter 24, "Employee Health and Benefits Management at EMC: An Information Driven Model for Engaged and Accountable Care," by David Dimond and Robert Morison.

Cheek notes that Aetna's analytic maturity is high (4+) on the Davenport-Harris Analytics Maturity Model relative to an industry average of stage 2. He describes the five primary services for its internal and external constituencies including 1) provider analytics to identify opportunities for outcome and cost improvements among physicians and hospitals, 2) plan-sponsor reporting for employers, 3) program evaluation of the ongoing effectiveness of care management programs, 4) custom informatics for "special projects," and 5) data warehousing.

In terms of the factors of analytics success, Cheek says that the most important are identifying the strategic drivers that offer the most demonstrable value from analytical enhancement, lodging the data warehouse with the informatics organization, and developing an internal analytics competency.

Diamond and Morison describe a different analytics focus, on the employee, and how the company promotes health for its workforce. The EMC vision is for employers to engage patients and providers, enable health awareness and literacy, influence health and lifestyle behaviors, and drive adoption of patient-centric technologies. The analytics to support the employee focus include an employee health portal, a personal health record, health risk assessments and incentives to be healthy, and the availability of related health management programs.

Life Science Companies

As Handelsman stated in Chapter 4, "Surveying the Analytical Landscape in Life Sciences Organizations," the life sciences industries are engaged in discovering, developing, and commercializing new therapies. The industry challenges are no less daunting than the rest of the health ecosystem. These include blockbuster drugs going off patent, health plan pressures to lower costs and demanding justification for the value of drugs, pricing pressures from generic drugs, long lasting investor caution following the Great Recession, the high cost and failure rate of clinical trials, and cost cutting that has cut into research innovation.

Analytics have been a core skill in the research and development discovery process and in determining value through comparative effectiveness studies. But, as in other aspects of healthcare, putting the research knowledge to use in improving clinical and business outcomes has lagged. There is great promise with the analytics of personalized medicine and the use of genomic to fill in gaps in products. And there is a renewed focus on customers that goes beyond direct to consumer advertising that can build loyalty and foster brand support.

In Chapter 25, "Commercial Analytics Relationships and Culture at Merck," Davenport reports on one specific life science industry analytical function, commercial analytics that focus on promotion and sales, at a major pharmaceutical firm, Merck. He concentrates on the "how" of making analytics work well for the business. It's all about decision support for the sales business. According to the business, the analytics function has been successful because the group members are "thought partners": they start with a full understanding of the business question and then marshal data to answer the questions, they are "field friendly" in translating findings into solutions, and they embed analytical results into software tools. Key questions about the future role of analytics are how to expand beyond the U.S. market and provide global support and how to create more collaboration among other analytics groups at Merck.

Notes

1. Micah Hartman et al., National Health Spending in 2011: Overall Growth Remains Low, but Some Payers and Services Show Signs of Acceleration, Health Affairs 32 (2013): 87-99.

2. U.S. Department of Health & Human Services, "HHS Secretary Kathleen Sebelius Announces Major Progress in Doctors, Hospital Use of Health Information Technology," February 12, 2012, www. hhs.gov/news/press/2012pres/02/20120217a.html.

18

Partners HealthCare System

Thomas H. Davenport

Partners HealthCare System (Partners) is the single largest provider of healthcare in the Boston area. It consists of 12 hospitals, with more than 7,000 affiliated physicians. It has 4 million outpatient visits and 160,000 inpatient admissions a year. Partners is a nonprofit organization with almost $8 billion in revenues, and it spends more than $1 billion per year on biomedical research. It is a major teaching affiliate of Harvard Medical School.

Partners is known as a "system," but it maintains substantial autonomy at each of its member hospitals. While some information systems (the electronic medical record, for example) are standardized across Partners, other systems and data, such as patient scheduling, are specific to particular hospitals. Analytical activities also take place both at the centralized Partners level and at individual hospitals such as Massachusetts General Hospital (MGH) and Brigham and Women's Hospital (usually described as "the Brigham"). In this chapter, both centralized and hospital-specific analytical resources are described. The focus for hospital-specific analytics is the two major teaching hospitals of Partners—MGH and the Brigham—although other Partners hospitals also have their own analytical capabilities and systems.

Centralized Data and Systems at Partners

The basis of any hospital's clinical information systems is the clinical data repository, which contains information on all patients, their conditions, and the treatments they have received. The inpatient clinical data repository for Partners was initially implemented at the Brigham during the 1980s. Richard Nesson, the Brigham and Women's CEO, and John Glaser, the hospital's chief information officer, initiated an outpatient electronic medical record (EMR) at the Brigham in 1989.[1] This EMR contributed outpatient data to the clinical data repository. The hospital was one of the first to embark on an EMR, though MGH had begun to develop one of the first full-function EMRs as early as 1976.

A clinical data repository provides the basic data about patients. Glaser and Nesson came to agree that in addition to a repository and an outpatient EMR, the Brigham—and Partners after 1994, when Glaser became its first CIO—needed facilities for doctors to input online orders for drugs, tests, and other treatments. Online ordering (called CPOE, or Computerized Provider Order Entry) would not only solve the time-honored problem of interpreting poor physician handwriting, but could also, if endowed with a bit of intelligence, check whether a particular order made sense or not for a particular patient. Did a prescribed drug comply with best-known medical practice, and did the patient have any adverse reactions in the past to it? Had the same test been prescribed six times before with no apparent benefit? Was the specialist to whom a patient was being referred covered by his or her health plan? With this type of medical and administrative knowledge built into the system, dangerous and time-consuming errors could be prevented. The Brigham embarked on its CPOE system in 1989.

Nesson and Glaser knew that there were other approaches to reducing medical error than CPOE. Some provider institutions, such as Intermountain Healthcare in Utah, were focused on close adherence by physicians to well-established medical protocols. Others, like Kaiser Permanente in California and the Cleveland Clinic, combined insurance and medical practices in ways that incented all providers to work jointly on behalf of patients. Nesson and Glaser admired those approaches, but felt that their impact would be less in an academic medical center such as Partners, where physicians were somewhat autonomous, and individual departments prided themselves on their separate reputations for research and practice innovations. Common, intelligent systems seemed like the best way to improve patient care at Partners.

In 1994, when the Brigham and Mass General combined as Partners HealthCare System, there was still considerable autonomy for individual hospitals in the combined organization. However, from the onset of the merger, the two hospitals agreed to use a common outpatient EMR called the longitudinal medical record (LMR) and a CPOE system, both of which were developed at the Brigham. This was powerful testimony in favor of the LMR and CPOE systems, since there was considerable rivalry between the two hospitals, and Mass General had its own EMR.

Perhaps the greatest challenge was in getting the extended network of Partners-affiliated physicians up on the LMR and CPOE. The physician network of more than 6,000 practicing generalist and specialist physician groups was scattered around the Boston metropolitan area, and often operated out of their own private offices. Many lacked the IT or telecom infrastructures to implement the systems on their own, and implementation of an outpatient EMR cost about $25,000 per physician. Yet full use of the system across Partners-affiliated providers was critical to a seamless patient experience across the organization.

Glaser and the Partners information systems (IS) organization worked diligently to spread the LMR and CPOE to the growing number of Partners hospitals and to Partners-affiliated physicians and medical practices. To assist in bringing physicians outside the hospitals on board, Partners negotiated payment schedules with insurance companies that rewarded physicians for supplying the kind of information available from the LMR and CPOE. By 2007, 90% of Partners-affiliated physicians were using the systems, and by 2009, 100% were. By 2009, more than 1,000 orders per hour were being entered through the CPOE system across Partners.

The combination of the LMR and the CPOE proved to be a powerful one in helping to avoid medical error. Adverse drug events, or the use of the wrong drug for the condition or one that caused an allergic reaction in the patient, typically were encountered by about 14 of every 1,000 inpatients. At the Brigham before LMR and CPOE, the number was about 11. After the widespread implementation of these systems at Brigham and Women's, there were just above five adverse drug events per 1,000 inpatients—a 55% reduction.

Managing Clinical Informatics and Knowledge at Partners

The Clinical Informatics Research & Development (CIRD) group, headed by Blackford Middleton, is one of the key centralized resources for healthcare analytics at Partners. Many of CIRD's staff, like Middleton, have multiple advanced degrees; Middleton has an MD, a Master of Public Health degree, and a Master of Science in Health Services Research.

The mission of CIRD is

> to improve the quality and efficiency of care for patients at Partners HealthCare System by assuring that the most advanced current knowledge about medical informatics (clinical computing) is incorporated into clinical information systems at Partners HealthCare.[2]

CIRD is part of the Partners IS organization. It was CIRD's role to help create the strategy for how Partners used information systems in patient care, and to develop both production systems capabilities and pilot projects that employ informatics and analytics. CIRD's work had played a substantial role in making Partners a worldwide leader in the use of data, analysis, and computerized knowledge to improve patient care. CIRD also has had several projects funded by U.S. government health agencies to adapt some of the same tools and approaches it developed for Partners to the broader healthcare system.

One key function of CIRD was to manage clinical knowledge, and translate healthcare research findings into daily medical practice at Partners. In addition to facilitating adoption of the LMR and CPOE, Partners faced a major challenge in getting control of the clinical knowledge that was made available to care providers through these and other systems. The "intelligent CPOE" strategy demanded that knowledge be online, accessible, and easily updated so that it could be referenced by and presented to care providers in real-time interactions with patients. There were, of course, a variety of other online knowledge tools, such as medical literature searching, available to Partners personnel; in total they were referred to as the "Partners

Handbook." At one point after use of the CPOE had become wide-spread at Brigham and Women's, a comparison was made between online usage of the Handbook and usage of the knowledge base from order entry. There were more than 13,000 daily accesses through the CPOE system at the Brigham alone, and only 3,000 daily accesses of the Handbook by all Partners personnel at all hospitals. There-fore, there was an ongoing effort to ensure that as much high-quality knowledge as possible made it into the CPOE.

The problem with knowledge at Partners was not that there wasn't enough of it; indeed, the various hospitals, labs, departments, and individuals were overflowing with knowledge. The problem was how to manage it. At one point, Tonya Hongsermeier, a physician with an MBA degree who was charged with managing knowledge at Partners, counted the number of places around Partners where there was some form of rule-based knowledge about clinical practice that was not centrally managed. She found about 23,000 of them. The knowledge was contained in a variety of formats: paper documents, computer "screen shots," process flow diagrams, references, and data or reports on clinical outcomes—all in a variety of locations, and only rarely shared.

Hongsermeier set out to create a "knowledge engineering and management" factory that would capture the knowledge at Partners, put it in a common format and central repository, and make it available for CPOE and other online systems. This required not only a new computer system for holding the thousands of rules that consti-tuted the knowledge, but an extensive human system for gathering, certifying, and maintaining the knowledge. It consisted of the follow-ing roles and organizations:

- A set of committees of senior physicians who oversaw clinical practice in various areas, such as the Partners Drug Therapy Committee, which reviewed and sanctioned the knowledge as correct or best known practice
- A group of subject matter experts who, using online collabora-tion systems, debated and refined knowledge such as the best drug for treating high cholesterol under various conditions, or the best treatment protocol for diabetes patients

- A cadre of "knowledge editors" who took the approved knowledge from these groups and put it into a rule-based form that would be accepted by the online knowledge repository

High Performance Medicine at Partners

Glaser and Partners IS had always had the support of senior Partners executives, but for the most part their involvement in the activities designed to build Partners' informatics and analytics capabilities was limited to some of the hospitals and those physician practices that wanted to be on the leading edge. Then Jim Mongan moved from being president of MGH (a role he had occupied since 1996, shortly after the creation of Partners) to being CEO of Partners overall in January 2003. Not since Dick Nesson had Glaser had such a strong partner in the executive suite.

Mongan had come to appreciate the value of the LMR and CPOE, and other clinical systems, while he headed Mass General. But when he came into the Partners CEO role, with responsibility over a variety of diverse and autonomous institutions, he began to view it differently. Mongan said:

> So when I was preparing to make the move to Partners, I began to think about what makes a health system. One of the keys that would unite us was the electronic record. I saw it as the connective tissue, the thing we had in common, that could help us get a handle on utilization, quality, and other issues.

Together Mongan and Glaser agreed that while Partners already had strong clinical systems and knowledge management compared to other institutions, a number of weaknesses still needed to be addressed (most importantly that the systems were not universally used across Partners care settings), and steps needed to be taken to get to the next level of capability. Working with other clinical leaders at Partners, they began to flesh out the vision for what came to be known as the High Performance Medicine (HPM) initiative, which took place between 2003 and 2009.

Glaser commented on the process the team followed to specify the details of the HPM initiative:

Shortly after he took the reins at Partners, however, Jim had a clear idea on where he wanted this to go. To help refine that vision, several of us went on a road trip, to learn from other highly integrated health systems such as Kaiser, Intermountain Healthcare, and the Veterans Administration about ways we might bring the components of our system closer together.

Mongan concluded:

We also were working with a core team of 15-20 clinical leaders and eventually came up with a list of seven or eight initiatives, which then needed to be prioritized. We did a "Survivor"-style voting process, to determine which initiatives to "kick off the island." That narrowed down the list to five Signature Initiatives.

The five initiatives consisted of the following specific programs, each of which was addressed by its own team:

- **Creating an IT infrastructure**—Much of the initial work of this program had already been done; it consisted of the LMR and the CPOE, which was extended to the other hospitals and physician practices in the Partners network and maintained. This project also addressed patient data quality reporting, further enhancement of knowledge management processes, and a patient data portal to give patients access to their own health information.

- **Enhancing patient safety**—The team addressing patient safety issues focused on four specific projects: 1) providing decision support about what medications to administer in several key areas, including renal and geriatric dosing; 2) communicating "clinically significant test results," particularly to physicians after their patients have left the hospital; 3) ensuring effective flow of information during patient care transitions and handoffs in hospitals and after discharge; 4) providing better decision support, patient education, and best practices and metrics for anticoagulation management.

- **Uniform high quality**—This team addressed quality improvement in the specific domains of hospital-based cardiac care, pneumonia, diabetes care, and smoking cessation; it employed both registries and decision support tools to do so.

- **Chronic disease management**—The team addressing disease management focused on prevention of hospital admission by identifying Partners patients who were at highest risk for hospitalization, and then developed health coaching programs to address patients with high levels of need, for example, heart failure patients; the team also pulled together a new database of information about patient wishes about end-of-life decisions.

- **Clinical resource management**—At Jim Mongan's suggestion, this team focused on how to lower the usage of high-cost drugs and high-cost imaging services; it employed both "low-tech" methods (e.g., chart reviews) and "high-tech" approaches (e.g., a data warehouse making transparent physicians' imaging behaviors relative to peers) to begin to make use of scarce resources more efficiently.

Overall, Partners spent about $100 million on HPM and related clinical systems initiatives, most of which were ultimately paid for by the Partners hospitals and physician practices that used them. To track progress, a Partners-wide report, called the HPM Close, was developed that shows current and trend performance on the achievement of quality, efficiency, and structural goals. The report was published quarterly to ensure timely feedback for measuring performance and supporting accountability across Partners.

New Analytical Challenges for Partners

Partners had made substantial progress on many of the basic approaches to clinical analytics, but there were many other areas at the intersection of health and analytics that it could still address. One was the area of personalized genetic medicine—the idea that patients would someday receive specific therapies based on their genomic, proteomic, and metabolic information. Partners had created the i2b2 (Informatics for Integrating Biology and the Bedside), a National

Center for Biomedical Computing that was funded by the National Institutes of Health. John Glaser was co-director of i2b2 and developed the IT infrastructure for the Partners Center for Personalized Genetic Medicine. One of the many issues these efforts addressed in personalized genetic medicine was how relevant genetic information would be included in the LMR.

Partners was also attempting to use clinical information for postmarket surveillance—the identification of problems with drugs and medical devices in patients after they have been released to the market. Some Partners researchers had identified dangerous side effects from certain drugs through analysis of LMR data. Specifically, research scientist John Brownstein's analyses suggested that the level of patients with heart attack admissions to Mass General and the Brigham had increased 18% beginning in 2001 and returned to its baseline level in 2004, which coincided with the timeframe for the beginning and end of Vioxx prescriptions. Thus far the identification of problems had taken place only after researchers from other institutions had identified them, but Partners executives believed it had the ability to identify them at an earlier stage. The institution was collaborating with the Food and Drug Administration and the Department of Defense to accelerate the surveillance process. John Glaser noted:

> I don't know that we'll get as much specificity as might be needed to really challenge whether a drug ought to be in a market, but I also think it's fairly clear that you can be much faster and involve much fewer funds, frankly, to do what we would call the "canary in the mine" approach.[3]

Partners was also focused on the use of communications technologies to improve patient care. Its Center for Connected Health, headed by Dr. Joe Kvedar, developed one of the first physician-to-physician online consultation services in an academic medical setting. The Center was also exploring combinations of remote monitoring technologies, sensors (for example, pill boxes that know whether today's dosage has been taken), and online communications and intelligence to improve patient adherence to medication regimes, engagement in personal health, and clinical outcomes.

In the clinical knowledge management area, Partners had done an impressive job of organizing and maintaining the many rules and knowledge bases that informed its "intelligent" CPOE system. However, it was apparent to Glaser, Blackford Middleton, and Tonya Hongsermeier—and her successor as head of knowledge management, Roberto Rocha—that it made little sense for each medical institution to develop its own knowledge base. Therefore, Partners was actively engaged in helping other institutions with the management of clinical knowledge. Middleton (the principal investigator), Hongsermeier, Rocha, and at least 13 other Partners employees were involved in a major Clinical Decision Support Consortium project funded by the U.S. Agency for Healthcare Research and Quality. The consortium involved a variety of other research institutions and healthcare companies, and was primarily focused on finding ways to make clinical knowledge widely available to healthcare providers through EMR and CPOE systems furnished by leading vendors.

Despite all these advances, not all Partners executives and physicians had fully bought into the vision of using smart information systems to improve patient care. Some found, for example, the LMR and CPOE to be invasive in the relationship of doctor and patient. A senior cardiologist at Brigham and Women's, for example, argued in an interview [with the author] that:

> I have a problem with the algorithmic approach to medicine. People end up making rote decisions that don't fit the patient, and it can also be medically quite wasteful. I don't have any choice here if I want to write prescriptions—virtually all of them are done online. But I must say that I am getting alert fatigue. Every time I write a prescription for nitroglycerine, I am given an alert that asks me to ensure that my patient isn't on Viagra. Don't you think I know that at this point? As for online treatment guidelines, I believe in them up to a point. But once something is in computerized guidelines it's sacrosanct, whether or not the data are legitimate. Recommendations should be given with notification of how certain we are about them.... Maybe these things are more useful to some doctors than others. If you're in a subspecialty like cardiology

you know it very well. But if you are an internist, you may have shallow knowledge, because you have to cover a wide variety of medical issues.

Many of the people involved in developing computer systems for patient care at Partners regarded these as valid concerns. "Alert fatigue," for example, had been recognized as a problem within Blackford Middleton's group for several years. They had tried to eliminate the more obvious alerts, and to make changes in the system to allow physicians to modify the types of alerts they received. There was a difficult line to draw, however, between saving physician attention and saving lives.

Centralized Business Analytics at Partners

While much of the centralized analytical activity at Partners has been on the clinical side, the organization is also making progress on business analytics. The primary focus of these efforts is on financial reporting and analysis.

For several years, for example, Partners has employed an external "software as a service" tool to provide reporting on the organization's revenue cycle. It has also developed several customized analytics applications in the areas of cash management, underpayments, bad debt reserves, and charge capture. These activities primarily took place in the Partners Revenue Finance function.

The Partners Information Systems organization is also increasing its focus on administrative and financial analytics. It is putting in place Compass, a common billing and administrative system, at all Partners hospitals. At the same time, Partners has created a set of standard processes for collecting, defining, and modifying financial and administrative data. Further, as one article put it:

> At Partners, John Stone, corporate director for financial and administrative systems, is developing a corporate center of business analytics and business intelligence. Some 12 to 14 financial executives will oversee the center, define Partners'

strategy for data management, and determine data-related budget priorities. "Our analysts spend the majority of their time gathering, cleaning, and scrubbing administrative data and less time providing value-added analytics and insight into what the data is saying," says Stone. "We want to flip that equation so our analysts are spending more time producing a story that goes along with the data."[4]

Hospital-Specific Analytical Activities— Massachusetts General Hospital

MGH, because it was a highly research-driven institution, had long focused primarily on clinical research and the resulting clinical informatics and analytics. In addition to the LMR and CPOE systems used by Partners overall, MGH researchers and staff have developed a number of IT tools to analyze and search clinical data, one of which was a tool that searched across multiple enterprise clinical systems, including the LMR.

While historically, the research, clinical, information systems, and the analytically focused business arms of MGH tended to operate in stove pipes, the challenges of an evolving healthcare landscape have forced a change in that paradigm. For instance, a strong current focus within MGH is on how to achieve federal "meaningful use" reimbursement for the organization's expenditures on EMR. Because achieving meaningful use objectives is predicated on a high level of coordination among information systems, the physicians, and business intelligence, people like David Y. Ting, the associate medical director for Information Systems for MGH and Massachusetts General Physicians Organization, and Chris Hutchins, the director of Finance Systems and deputy CIO, are beginning to collaborate extensively.

The HITECH/ARRA criteria for Stage 1 EMR meaningful use prescribe 25 specific objectives to incentivize providers to adopt and use electronic health records.[5]

To raise the level of EMR use by all its providers, as well as to provide resources for the work needed to achieve that level, MGH has arrived at a novel funds distribution model. They determined that the physicians organization will reserve a portion of the pool of $44,000 per physician toward IT and analytics infrastructure, then distribute the remaining incentive payment across all providers, proportional to the amount of data a particular physician is charged with entering. An internal quality incentive program would serve as the distribution mechanism. So, for example, if you recorded demographics, vital signs, and smoking status for the requisite number of patients, you would receive 30% of the per-physician payment from the pool. If you fulfilled all ten quality measures, you would receive 100% of the payment from the pool. This encourages all physicians to contribute to the meaningful use program, but it also means that no physicians will receive the full amount of $44,000. The incentive from the federal government is up to $44,000 for each eligible provider who fulfills the meaningful use criteria. MGH has examined the objectives and broken them down into ten major pieces of patient data that physicians need to record in the EMR. However, many are not relevant for all of its physicians. For example, a primary care physician would logically enter such data as demographics, vital signs, and smoking status, but these would be less relevant for certain specialists to enter.

Clearly, such a complex quality incentive model requires an unprecedented level of analytics. Currently, Ting, Hutchins, and others at MGH are working to map the myriad clinical and finance data sources that are scattered among individual departments, exist at a hospital site level, or exist at the Partners enterprise level. Simultaneously, they must negotiate data governance agreements even among other Partners entities, to ensure that the requisite data feeds from sources within Partners and pertaining to MGH, but stored outside MGH's physical data warehouses, are available for MGH analytics purposes.

MGH has some experience with reimbursement metrics based on physician behaviors, having used them in Partners Community HealthCare, Inc. (PCHI), its physician network in the Boston area. Physician incentives have been provided through PCHI on the basis of admission rates, cost-effective use of pharmacy and imaging services, and screening for particular diseases and conditions, such as diabetes. This was also the mechanism used to encourage the adoption of the LMR and CPOE systems by physicians. But MGH, like other providers, struggles with developing clear and transparent metrics across the institution that can help to drive awareness and new behaviors. If MGH could create broadly accessible metrics on individual physicians' frequency of prescribing generic drugs, for example, it would undoubtedly drive MGH's competitive physicians to excel in the rankings.

On the business side, MGH is trying to develop a broad set of capabilities in business intelligence and analytics. A Business Intelligence/ Performance Management group has recently been created under the direction of Chris Hutchins, deputy CIO and director of finance systems for the Mass General Physicians Organization (MGPO). The group is generating reports on such financial and administrative topics as

- Billing efficiency, claims adjudication, rejection rates, and times to resolve billing accounts, both at MGH overall and across practices
- Improving patient access, average wait times to see a physician, and cancellation and no show rates
- Employer attrition as an MGH customer

MGH is also working with CMS on the Physician Quality Reporting Initiative. To combine all these measures in a meaningful fashion, MGPO is also working on a balanced scorecard.[6]

While the current analytical activity is largely around reporting, Hutchins plans to develop more capabilities around alerts, exception reporting, and predictive models. The MGH Physicians Organization is implementing capabilities for statistical and predictive analytics that would be applied to several topics. For example, one key area in which better prediction would be useful involves patient volume. They are also pursuing more general models that would predict shifts in business over time. At the moment, however, Hutchins feels that the scorecard is still early in its development and current efforts are focused on identifying leading indicators.

Hospital-Specific Analytical Activities— Brigham and Women's Hospital

Like MGH, the Brigham's analytical activities in the past have been largely focused on clinical research. Today it is also addressing much of the same business, operational, and meaningful use issues that MGH is. Many of the analytical activities at the Brigham are pursued by the Center for Clinical Excellence (CCE), which was founded by Dr. Michael Gustafson in 2001. The center has five functionally interrelated sections, including

- Quality programs
- Patient safety
- Performance improvement
- Decision support systems (including all internal and external data management and reporting activities)
- Analysis and planning (which oversees business plan development, ROI assessments for major investments, cost benchmarking, asset utilization reporting, and support for strategic planning)

The CCE has close working relationships with the Brigham's CFO and finance organizations, the Brigham's information systems organization, the Partners Business Development and Planning function, and other centers and medical departments at the Brigham.

One major difference between the Brigham and MGH (and most other hospitals, for that matter) is that the Brigham established a balanced scorecard beginning in 2000. It was based on a well-established cultural orientation to operational and quality metrics throughout the hospital. Richard Nesson, the Brigham CEO who had partnered with CIO John Glaser to introduce the LMR and CPOE systems, was also a strong advocate of information-driven decision making on both the clinical and business sides of the hospital. The original systems that Nesson and Glaser had established also incorporated a reporting tool called EX, and a data warehouse called CHASE (Computerized Hospital Analysis System for Efficiency). The analyses and data from these systems formed the core of the Brigham's balanced scorecard.

Before an effective scorecard could be developed, the Brigham had to undertake considerable work on data definitions and management. One analysis discovered, for example, that there were five different definitions of the length of a patient stay circulating in 11 different reports. The chief medical officer at the time, Dr. Andy Whittemore, and the CCE's Dr. Gustafson, a surgeon who had just taken on quality measurement issues at the Brigham, addressed these data issues with a senior executive steering committee and decided to present the data in an easy-to-digest scorecard.

Under the ongoing management of the CCE, the scorecard contains a variety of financial, operational, and clinical metrics from across the hospital. The choice of metrics is driven by a "strategy map"[7] specifying the relationships among key variables that drive the performance of the hospital (see Figure 18.1). Unlike most corporate strategy maps, financial performance variables are at the bottom of the map rather than the top. In the scorecard itself, there are more than 50 specific measures in the hospital-wide scorecard, and more detailed scorecards for particular departments, such as Nursing and Surgery. The scorecard has also been extended to Faulkner Hospital, a Partners institution that is managed jointly with the Brigham.

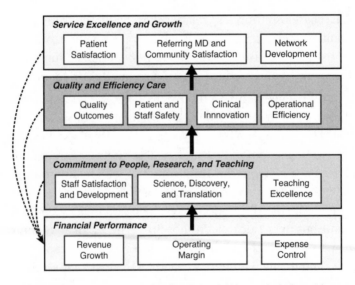

Figure 18.1 Strategy map for Brigham & Women's balanced scorecard

Dr. Gary Gottlieb, the Brigham president from 1992 to 2009, was the most aggressive user of the scorecard. He noted:

> I review the balanced scorecard on a regular basis, because there is specific data that is of interest to me. There are key metrics I examine for trends and if they develop, then I analyze the data to better understand what is going right or wrong. It is one view, but an important one of our hospital. I can look at the balanced scorecard and get information in another way, from a different perspective than I can when I'm making rounds on a hospital unit, or sitting in the meeting with chiefs.[8]

Gottlieb left the Brigham CEO role to become the CEO of Partners overall in 2010. One of the primary initiatives in his new Partners role is to expand the degree of common systems throughout Partners, so that there can be common data and analytics throughout the organization. Perhaps one day all of Partners HealthCare System will be managed through one scorecard.

Notes

1. This and other details of the Partners LMR/CPOE systems are derived from Richard Kesner, "Partners Healthcare System: Transforming Healthcare Services Delivery Through Information Management," Ivey School of Business Case Study (2009).

2. "CIRD, Clinical Informatics Research & Development," http://www.partners.org/cird/.

3. PricewaterhouseCoopers, "Partners HealthCare: Using EHR Data for Post-market Surveillance of Drugs" (2009). http://pwchealth.com/cgi-local/hregister.cgi/reg/partners_healthcare_case_study.pdf.

4. Healthcare Financial Management Association, "Developing a Meaningful EHR," http://www.hfma.org/Publications/Leadership-Publication/Archives/Special-Reports/Spring-2010/Developing-a-Meaningful-EHR/, Part 3 of "Leadership Spring-Summer 2010 Report: Collaborating for Results."

5. The 25 meaningful use criteria are described in "Eligible Provider: 'Meaningful Use' Criteria," by Jack Beaudoin, Healthcare IT News, December 30, 2009, http://www.healthcareitnews.com/news/eligible-provider-meaningful-use-criteria.

6. Robert S. Kaplan and David P. Norton, "The Balanced Scorecard: Measures that Drive Performance," *Harvard Business Review* (January – February 1992).

7. Robert S. Kaplan and David P. Norton, "Having Trouble With Your Strategy? Then Map It," *Harvard Business Review* (September – October, 2000).

8. Ibid.

19

Catholic Health Initiatives

Deborah Bulger and Evon Holladay

Healthcare organizations sometimes struggle with managing the volumes of data they produce—from financial, clinical, operational systems, and processes. Yet the ability to manage data and transform it into meaningful information yields significant returns to an organization's business performance. A recent report by Yonek et al. on the characteristics of high-performing healthcare organizations cited several best practices including:

- Establish a systemwide strategic plan with measurable goals and track progress toward achieving them with system performance dashboards.

- Create alignment across the health system with goals and incentives.

- Leverage data and measurement across the organization by, among other things, frequently sharing dashboards and national benchmarks with hospital leaders and staff to identify areas in need of improvement and taking immediate actions to get back on track.[1]

To be effective, information must reach the people charged with improving performance, and must reach them in a timely and appropriate fashion. At the enterprise level, a global, measurable strategy will set direction for the organization. At an operational level, measures that support financial and capacity related activities should be readily available. And for measures related to patient care activities, reporting must reach caregivers in real-time. The challenge: How does a healthcare organization translate data into actionable information for every stakeholder across the enterprise?

About the Organization

Catholic Health Initiatives (CHI) is a national nonprofit health organization with headquarters in Denver. It is a faith-based system that includes 73 hospitals; more than 400 physician practices; 40 long-term care, assisted- and residential-living facilities; a home health agency; and two community health-services organizations in 19 states. Together, its facilities serve more than 60 rural and urban communities and provided nearly $500 million in community benefit in the 2010 fiscal year, including services for the poor, free clinics, education, and research. With approximately 70,000 employees and annual revenues of more than $8 billion, CHI ranks as the nation's third-largest Catholic healthcare system. It is ever moving toward its vision of Catholic healthcare as a vibrant ministry, ready to provide compassionate care of the body, mind, and spirit through the twenty-first century and beyond.

Business intelligence (BI) is a relatively new function for CHI. It is responsible for providing a historical, current, and predictive view of business operations through an enterprise data warehouse. CHI's patient data warehouse provides information for strategic reporting and core measures of regulatory compliance. The department is partnering with leadership across CHI to develop metric standards and define key performance indicators and best practices benchmarks. The goal is to reduce latency in decision making by having information readily available.

Current Situation

Because of the breadth and depth of services provided and geography covered, CHI represents a microcosm of the U.S. healthcare delivery system. The distributed nature of the organization, disparity of systems, and the sheer magnitude of data produced across the enterprise all contribute to the complexity of data standardization. For instance, CHI uses multiple vendors across the enterprise for hospital information and acute care billing systems, clinical decision support, compensation, revenue management, enterprise resource planning (ERP), productivity, and so on. In physician practices alone there are 14 different vendor solutions. To achieve an enterprise reporting model, CHI is leveraging commercially available tools for BI, data marts, extract/ transform and load (ETL), and enterprise data warehousing.

Like Yonek et al., CHI recognizes a need for a strategic alignment of technology, information, and stewardship if the organization is to move up the analytics maturity curve. In the book *Analytics at Work: Smarter Decisions, Better Results*, Davenport et al., explains how organizations can use data and analysis to make better decisions.[2] Aimed at a broad, multidisciplinary audience, it speaks to employees across their organizations who want to know where they stand now and what they need to do to become more analytical over time. The DELTA[3] model outlines five key components for deploying and succeeding with analytical initiatives:

- D for accessible, high-quality **data**
- E for an **enterprise** orientation
- L for analytical **leadership**
- T for strategic goals or **targets**
- A for **analytical** talent

It is through this model that we describe CHI's journey toward enterprise intelligence.

Data

Data are the foundation for analytics, and CHI recognizes that managing data across the enterprise requires discipline. The organization has identified three critical steps to establishing an enterprise data model.

Selection of the Standard

At CHI, the lack of consistent definitions has, at times, led to false assumptions about an individual organization's performance, creating barriers to an enterprise approach. To first agree on relatively simple definitions, such as whether to include day surgery or lab visits in volume measures, makes it easier to tackle more complex definitions such as adjusted patient days. CHI is designing this model operationally through its governance structure (discussed later in this chapter) and has started the standardization process with acute care, to be followed with physician practices and home care.

Of particular importance are the partnerships CHI has developed with its software vendors to standardize naming conventions within their products. This helps to ensure definitions are consistent even outside the enterprise data warehouse.

Implementation

Once standards are determined, the organization needs to ensure that they are implemented. CHI plans to use data governance to understand business requirements, design data definitions, develop and test metrics, and ensure effective implementation.

A change control process that was started in BI will be adopted across all CHI reporting systems. This process begins with a gap analysis to evaluate new data definitions or changes to calculations and determines when those changes will be activated. To ensure accountability, CHI uses a RACI[4] matrix to assign people who are responsible, accountable, consulted, and informed for activities and decisions that impact implementation.

Ongoing Monitoring

It is critical to monitor data continuously to ensure its integrity, identify new measures, and assess the information behaviors of people who gather and use it. Administrative data are highly standardized, but as new metrics demand a higher level of data accuracy, it is necessary to train people to collect data correctly. One example at CHI is the capture of patient race and ethnicity data. These are key data elements used for monitoring underserved populations and mitigating disparities in care, a critical component of health reform. There are national standards for this—no need to recreate the process—but the data are not always captured accurately upon admission. A high percentage of patient records at CHI were listed as "unable to determine." Rather than take a punitive approach, the organization chose to address the issue as a function of behavior and ongoing process. Once admission personnel understood how these data were used and the importance to patient care, capture rates improved.

Data Management in Action

An example of a decision that impacts implementation is determining whether data must go to the warehouse for further normalization, aggregation, or modeling, or if they can be viewed directly through BI tools in transaction systems or other reporting systems like cost accounting. The decision is based on the frequency of data needs (e.g., real-time patient census supported by a single variable) and the complexity of analysis, such as projected payments requiring statistical models inherent in the data warehouse.

Enterprise

CHI aggregates financial and operational data supplied by each entity. However, it has not historically provided an enterprise reporting methodology with a standard taxonomy for comparing key business practices. As the healthcare delivery model becomes more disparate—acute, ambulatory, home care, long-term care—it creates some interesting challenges for comparisons at an enterprise level. For example, the acceptable operating margin may be very different for long-term care than acute care, so it is important to understand the reason for the difference and recognize the unique contributions of each care setting. CHI's objective is to view the two holistically, align around a model of shared accountability, and recognize that long-term care is an equal partner in the delivery model. As long-term care centers are added to the organization—40 centers at this writing—they are treated as part of the enterprise rather than independent entities. CHI's mission is to achieve the best possible care for the community by ensuring appropriate handoffs and measuring effectiveness across the continuum.

Leadership

Becoming accountable for the care of a broad and diverse patient population means that strategic decisions must be made based on reliable data. CHI has structured its information management model to support an enterprise strategy for decision making (see Figure 19.1).

Figure 19.1 Simplified graphic of Catholic Health Initiatives data governance data structure

In this model, senior leaders drive a top down directive for corporate alignment of strategic metrics with measurable action plans.

At an operational level, the information management council is a multidisciplinary group of individuals selected by their senior operational leaders, representing all regions of CHI. The objective of this group is to define priority solutions that will provide stakeholders with timely data so they can make better decisions based on information rather than anecdote.

CHI's data governance structure is composed of functional vice presidents from across the enterprise. Their goal is to create accountability for standardized information that aligns with functional and operational priorities and external best practices. In addition, smaller workgroups aimed at specific business issues are designed to create an environment where participants can build relationships, have difficult conversations, and solve problems.

Target or Goal

Standardized data definitions allow for meaningful comparisons. CHI is large enough to benchmark results both internally and externally. This allows the organization to define current achievement and set aspirational goals. CHI's enterprise measurement dashboard is completely transparent, allowing each facility to view every other facility's results. These open comparisons enable organizations to see who is doing better on various metrics and to share the possibilities. It is acceptable for entities to be different as long as they explore those differences and are driven to improve. BI's goal is to create energy and power by providing information that facilitates learning.

To drive that energy, CHI is shifting the analytics paradigm to include more exploratory analytics that empower local organizations to ask and answer difficult questions about their results. CHI recognizes two important aspects of this applied learning model:

- **Access to truly comparative information**—Organizations can "drill" to comparisons on the dashboard at the market level and across all facilities with confidence in the results.

- **Collaboration**—At a functional level, organizations know each other and ask questions to understand the results and create shared practices. This approach creates an internal consultancy of shared practices based on long-term relationships that can be leveraged as needed.

By comparing outside its own organization, each entity can leverage the value of the collective enterprise knowledge.

Analysts

Analytical talent is necessary to "connect the dots" from data source to end results to provide critical insights. Technology enables analysis, but it is human capital that most benefits organizations that compete based on analytics. Three levels of analysts are described by Davenport et al.[5]

- Senior management sets the tone for analytical culture and makes decisions.
- Professional analysts gather and analyze data and report results to decision makers.
- Analytics "amateurs" use the outputs to perform their jobs.

This is an area where CHI, like many healthcare organizations, continues to learn. At this point in the journey, CHI is on the cusp of Stage 3—still dependent on localized analytics but aspiring to become an analytical organization (see Figure 19.2) with leaders who are setting the analytics tone of the organization. The next steps are to start aligning analysts—both professional and "amateur"—around a common understanding of data and measurement and to "get everyone out of their silos." As these capabilities mature, analysts will help build a framework of trust in both the data and shared practices that drive improvement.

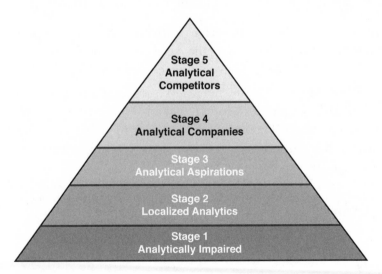

Figure 19.2 Five levels of analytics capability

Conclusion

The focus on coordination of care as an industry driver, the expansion of the care delivery model outside acute care, and pivotal leadership changes have created the "perfect storm" for CHI. As it moves into the next phase of analytics maturity, CHI will plan for additional milestones:

- Evaluate the concept of incentives tied to improving performance by measuring the "return on information" investment.
- Develop enterprise intelligence solutions that provide data in a more real-time manner. Ideally every stakeholder should have a common set of standardized information at her fingertips.
- Define an analytics roadmap. CHI has engaged consultants to help the organization accelerate deployment of high-value analytics.
- Find those true areas that need to improve by deploying more advanced analytics.
- Leverage meaningful use requirements as a baseline to create the next level of learning.

CHI views this process as a journey that promises to weave information into its business practices and deliver against predictable milestones in the future. The organization has made significant progress in the last 14 months as its culture has shifted to support an enterprise model, but there is still much work to do.

Notes

1. J. Yonek, S. Hines, and M. Joshi, *A Guide to Achieving High Performance in Multi-Hospital Health Systems*, Health Research & Educational Trust, March 2010, http://www.commonwealthfund. org/Content/Publications/Fund-Reports/2010/Mar/A-Guide-to-Achieving-High-Performance-in-MultiHospital-Health-Systems. aspx?utm_source=feedburner&utm_medium=feed&utm_campaign =Feed%3A+TheCommonwealthFund+%28The+Commonwealth+ Fund%29.

2. Thomas H. Davenport, Jeanne G. Harris, and Robert F. Morison, *Analytics at Work: Smarter Decisions, Better Results* (Harvard Business School Press, 2010).

3. Davenport et al. DELTA is also the Greek letter that signifies "change" in an equation.

4. RACI is an acronym that stands for responsible, accountable, consulted, and informed and deploys a matrix to assign. There are many references to RACI. This is one of them: http://www.projectsmart.co.uk/how-to-do-raci-charting-and-analysis.html.

5. Thomas Davenport and Jeanne G. Harris, *Competing on Analytics: The New Science of Winning* (Harvard Business School Press, 2007).

20

Analytics at the Veterans Health Administration

Thomas H. Davenport

The Veterans Health Administration (VHA), a unit of the U.S. Department of Veteran Affairs (VA), provides medical assistance to military veterans through 152 hospitals and medical centers, 784 outpatient clinics, and more than 100 long-term care facilities such as nursing homes. It serves a veteran population of more than 22 million and is the largest medical system in the United States.

The VA had some historical issues with quality of care for veterans, but for the last decade, it has performed well in that regard. Under its leader Dr. Kenneth Kizer, in the mid-1990s the VHA embarked on a major transformation in care quality and cost reduction. As one aspect of the transformation, the VA shifted resources from inpatient to outpatient care. At the same time, it decreased staffing while improving patient outcomes. For example, in only the four years from 1994 to 1998, the VA made the following changes in care programs:

- Closed 54% of acute care beds.
- Decreased bed-days by 62%.
- Decreased staffing by 11%.
- Increased the number of patients treated by 18%.
- Increased ambulatory visits by 35%.
- Instituted universal primary care.
- Reduced morbidity rates by 30%.

- Reduced mortality rates by 9%.
- Eliminated 72% of all forms.[1]

Veterans routinely rank the VA system as having better quality than other treatment alternatives, according to the American Customer Satisfaction Index. In 2008, the VA had a satisfaction rating of 85 for inpatient treatment, compared with 77 for private hospitals, and VA outpatient care outscored private hospital outpatient care by three points.[2]

The VA's EMR System and Related Analytics

The VA was one of the first large government care providers—or large providers of any type—to put a standard electronic medical record in place throughout the system. Originally known as the Decentralized Hospital Computer Program (DHCP), it was adopted in the 1980s. DHCP is still the core of the health information system in individual medical centers, though it has gained many new functions over the years. Renamed VistA (Veterans Health Information Systems and Technology Architecture) under Kizer in 1996, it was made available to other healthcare organizations under an open source arrangement. It included functionality such as wireless laptop access in patient rooms, bar coding of medications, electronic signatures for procedures, and access to online images. The VA had also recently added an online patient portal to VistA functionality. The portal reminded patients about allergies and medications, listed past and upcoming visits to VA facilities, and allowed e-mailed communications with care providers. Some patients had automated links between home monitoring devices and their VistA medical records.

VistA and other VA patient data were also increasingly being used for analytical purposes. In VistA itself, clinicians could create, for example, a chart of risk factors and medications to decide treatments. They could also search VistA records to find out, for example, if veterans were showing symptoms of diseases related to Agent Orange exposure.[3] When a VA hospital in Kansas City noticed an outbreak of a rare form of pneumonia among its patients, a quick search revealed that all the patients had been treated with a contaminated nasal spray.[4] In another example, VA researchers used VistA data to examine 12,000 medical records to explore treatment variations for diabetes across different VA doctors, hospitals, and clinics, and how patients fared under the different circumstances. The findings were then incorporated into clinical guidelines delivered by the VistA system. In the 1990s, the VA began using VistA data to identify underperforming and particularly successful surgical teams or hospital managers with regard to quality and safety benchmarks.[5]

In addition, the VA made available a variety of SAS datasets for analysis by researchers. These were extracts from a large patient care data warehouse using VistA data, and they addressed topics such as inpatient care and procedures, and outpatient visits and events. The patient records were normally anonymized.

The VA also maintained the Decision Support System, a managerial cost accounting system based on commercial software. It combined clinical and cost data to allow cost allocation to patient care products and services. Fully implemented by 1999, it allowed the VA to integrate expenses, workload, and patient utilization and outcomes. There were also data warehouses for each regional Veterans Integrated Services Network (VISN), and a Pharmacy Benefits Management Services database of all prescriptions issued by the VA.

To facilitate access to these tools by researchers and analysts, the VA maintained a VA Information Resource Center, an online portal that served as a guide to available research data, tools, and services.

All of these systems and analytical tools are employed throughout the VA. They are evidence that the VA is a leader in both clinical informatics and the performance improvements based on them. Some of the specific groups who use the data and perform the analyses are described in the following sections, along with some of their analytical activities.

Analytical Groups and Initiatives

In addition to the earlier VistA analytics efforts, many analytical initiatives involving patient care at the VA over the past several years have taken place in the office of the Under Secretary for Health for

Quality and Safety. That office includes three units that perform analytical work for the VA:

- **Quality and Safety Analytics Center (QSAC)**—This umbrella unit supports analysis and learning from the vast amount of patient data available from VistA. Not all of the data are useful for analytical purposes, so one of the tasks of QSAC is to determine which data elements are suitable for detailed analysis. QSAC has two specialized units under it: Inpatient Evaluation Center and Office of Productivity, Efficiency, and Staffing.

- **Inpatient Evaluation Center (IPEC)**—IPEC houses more than 20 quantitative analysts. It focuses on analysis of data on inpatient care to improve patient outcomes. It provides data and analysis to care providers and managers, focusing initially on intensive care units (ICUs), and later on acute care more generally. One of its first activities was to develop risk-adjusted metrics of patient outcomes that could be used throughout the VA. The risk adjustment method was based on data from an extraction program run at each medical center. The data were then analyzed to create reports that compared risk-adjusted mortality and length of stay to the medical center's adherence to process measures. ICU performance was compared against "average" and "best" performance benchmarks. ICU reports were segmented by specific patient groups such as the type of intensive care unit, severity of illness, or admission diagnosis, or procedure. Using a Web-based database application created and supported by the VA IPEC, hospitals could also track their hospital-acquired infections in the intensive care units.

  IPEC also supports the identification of other evidence-based practices to improve the care of veterans and their families. It addresses practices related to central line infections and ventilator-assisted pneumonia, and also focuses on practices that avoid urinary tract infections.

- **Office of Productivity, Efficiency, and Staffing (OPES)**— The OPES, which includes mathematicians and economists, undertook a variety of projects on the business side of VHA performance. It assesses such topics as clinical productivity, staffing levels, and overall efficiency. One key focus is the productivity of primary care physicians and some specialists, segmenting clinicians by teaching mission, practice setting, and patient complexity. Clinicians as well as VA system leaders are the primary audience for OPES analyses.

Analytics is also performed in other VA groups, although the purpose of the analyses is often research and publication (as it sometimes is in IPEC and OPES) more than changes in medical processes and treatment protocols. For example, the VA's Health Service Research and Development organization conducts rigorous research and publishes it in medical journals. There is also an outcomes analysis group in the VA's Surgery department, as well as other quantitative analysts in the Policy and Planning organization and a predictive modeling group in the Patient Treatment File organization.

The VA Office of Information and Technology also maintains a series of data warehouses (both a centralized "corporate" data warehouse and several regional warehouses), and a Business Intelligence Service Line to help with field operations information needs.

Quality Metrics and Dashboards

A key component of the 1990s care transformation at the VA involved a new framework by which to measure quality, and holding senior managers accountable for improvements in performance measures. The quality care framework includes morbidity rates, mortality rates, longevity (for example, one-year survival rates), functionality scores, and performance indicators. Throughout the VA, patient function is measured by a version of the SF-12 Health Survey. Performance indicators include a prevention index (for example, vaccination rates, cancer screening) and chronic disease care indices (such as hypertension control).

These metrics proliferated through multiple systems, displays, and dashboards. One of the early activities of the quality and safety analytics groups was to establish a single quality and safety dashboard for the VA. Called Links, it contains both process and outcome measures, with about seven mortality metrics for each facility.

In addition to Links, the VA's analytics groups also experimented with a variety of other dashboards and displays. Quality information, for example, is presented on statistical process control charts. Different medical facilities are compared on a "variable life adjusted display." The quality and efficiency levels of different facilities are compared in a "stochastic frontier analysis" displaying efficiency as a frontier of quality. Information in dashboards is often color-coded, and whenever possible the display shows trends and movement.

The quality and safety analytics groups also worked with new metrics, including medical center readmission rates, mental health readmissions, and 12 different ambulatory care conditions. If the 12 conditions are present, patients should not be admitted to medical centers. IPEC also developed measures of patient case severity and tracked whether particular facilities were admitting the types of cases that they were prepared to address.

Many of the metrics were presented using geographical comparisons. The researchers found a high degree of variation in quality and efficiency across the various VA facilities. The metrics and reports were intended to identify underperformers and best practices.

One key challenge at the VA is the amount of data in the organization. Analysts work to provide not just *more* information, but greater insight. For example, VA analysts generated a comprehensive Brief Analytical Review of quality and safety findings. It includes a variety of data sources, from internal quality data to peer reviews, surveys, and Joint Commission reports. All of the data go into a single report that "tells a story with data."

The goal of these analytical initiatives, of course, is to stimulate improvement in quality, safety, and efficiency—particularly in problematic facilities. In 2010 the VA began to post many of the measures on the Internet, including various death rates, intravenous line infection rates, ventilator-acquired pneumonia rates, and readmission rates. The VA mounted interventions for hospitals that fall into the bottom decile of national results, and some doctors and administrators can lose their jobs. One article mentioned:

> "The VA secretary pays attention to this," says William E. Duncan, the agency's associate deputy undersecretary for health quality and safety. "Unless people in the VA system

have an organizational death wish, they will pay attention to this, too."[6]

These efforts show clear payoffs. Central line infections, for example, were reduced by two-thirds. Similar reductions were achieved in the incidence of ventilator-associated pneumonia.

Possible Future Uses of Analytics at the VA

Much of the past work of analytics groups at the VA involved reporting of the organization's various metrics. However, analysts are beginning to focus on ways to predict and optimize important phenomena in the care of veterans.

For example, in 2011 the analytics groups at the VA, particularly the IPEC, were exploring the use of a neural network to predict the most likely high-risk sites. Four or five percent of facilities fall into that category each quarter, and analysts want to predict and address poor outcomes ahead of the problem. More broadly, the analytics groups at the VA want to get into greater use of predictive modeling and optimization.

In addition to inpatient and outpatient care, the VA's analytics groups are also beginning to address "fee care," or care provided by non-VA facilities and clinicians for a fee. The VA spends more than $4 billion each year in external fee-based care. In general, research found that VA care is less expensive than that provided externally for a fee, and the VA analysts wanted to learn who the outliers are in buying fee care, under what circumstances fee care is used, and how effective it is.

Many of the VA's future analytical activities likely involve analyses involving people and the human skills to do analytical decision making. One approach might involve exploring the relationships between a facility's employee attitudes and leadership behaviors, and health outcomes at those facilities. There is already some evidence that facilities with poor leadership scores and low psychological safety levels among employees have poor health outcomes. In the near future, the VA's analytical leaders hope to convert these research findings into intervention strategies.

There might also be future initiatives to nurture the analytical skills of VHA managers and clinicians. The quality and safety organization established an Analytics Academy, which offers quarterly training sessions around the country. Average attendance has grown from 25 to 80 per class, and these efforts might expand in the future. The new analytics organization also promises greater collaboration between analytics providers within the VHA on the key problems and decisions of the organization.

While the VA has made considerable strides with analytics and is clearly among the most aggressive users of clinical analytics in the United States, there is no complacency among analytical leaders. Pockets of the organization still resist evidence-based change, and analytical executives discuss helping the VHA overcome the "five stages of data grief." The goal, of course, is to get past any grief about poor performance and fix the problem. This attitude continues to serve the VA well as it moves toward a more analytical future.

Notes

1. Speech by Kenneth Kizer, "Reinventing Government-Provided Healthcare: The 'New VA,'" Leonard Davis Institute of Health Economics, University of Pennsylvania, 30 April 1999.

2. "Vets Loving Socialized Medicine Show Government Offers Savings," Bloomberg, October 2, 2009.

3. Gary Hicks, "A. O. Miner: Speeding Benefits to Vietnam Vets," *Vanguard*, U.S. Dept. of Veterans Affairs, Nov./Dec. 2010, p. 6.

4. Phillip Longman, "The Best Care Anywhere," *Washington Monthly*, Jan./Feb. 2005, online at http://www.washingtonmonthly.com/features/2005/0501.longman.html.

5. Phillip Longman, "Code Red," *Washington Monthly*, July-August 2009, online at http://www.washingtonmonthly.com/features/2009/0907.longman.html.

6. William M. Burton, "Data Spur Changes in VA Care," *The Wall Street Journal*, March 29, 2011.

21

The Health Service Data Warehouse Project at the Air Force Medical Service (AFMS)

Albert Bonnema and Jesus Zarate

The Air Force Medical Service (AFMS) works in close coordination with the Assistant Secretary of Defense for Health Affairs, the major air command surgeons, the Departments of the Army, Navy, and other government agencies to deliver medical services for more than two million eligible beneficiaries. Beneficiaries include active duty, family members, and retirees, during both peacetime and wartime.

The AFMS consists of approximately 38,000 officers, enlisted and civilian personnel, plus an additional 15,000 members assigned to the Air Force Reserves and the Air National Guard. The AFMS has an annual budget of approximately $5.4 billion and runs 75 military treatment facilities, including 16 hospitals and medical centers.

Vision and Mission

The AFMS's vision is to provide quality, world-class healthcare and health service support to eligible beneficiaries anywhere in the world at any time. The AFMS's mission is to provide seamless health service support to the United States Air Force (USAF) and combatant commanders. The AFMS assists in sustaining the performance, health, and fitness of every airman. It promotes and advocates for optimizing human performance for the war fighters, including the optimal integration of human capabilities with systems.

The AFMS operates and manages a worldwide healthcare system capable of responding to a full spectrum of health requirements. This ranges from providing care in forward-deployed locations to comprehensive preventive care.

The AFMS Office of the CIO

Under the direction of the AFMS Surgeon General, the AFMS Office of the CIO (OCIO) manages the strategic vision, implementation, delivery, and interdependence of all AFMS information management (IM) and information technology (IT) programs, including clinical information systems and healthcare informatics. Informatics responsibilities include portfolio and program management, budgeting, and stakeholder leadership, as well as oversight of project execution.

The OCIO addresses daily tactical challenges born of multiple concurrent projects to modernize the AFMS's information management and information technology. The OCIO is focused on creating next-generation capabilities and infrastructure while ensuring business continuity.

Efforts are currently under way to reshape the OCIO, bringing in new skill sets and creating an infrastructure and architecture that can support the AFMS for the long term.

AFMS's Modernization Challenges

The AFMS already meets many "meaningful use" mandates and has done so for years. But to operate more efficiently and cost effectively, the AFMS's IM/IT infrastructure must become modernized and integrated. Among AFMS's key challenges are

- **Data integration**—The AFMS has had electronic capabilities since the 1990s and has used electronic health records (EHRs) since the early 2000s. The amount of data that the AFMS has is astounding. Currently the AFMS receives 400 different data feeds and near real- time data from 101 sites around the world.

These data feeds include roughly eight million transactions each day.

However, data acquisition and integration has been developed organically and on a solution-by-solution basis, without alignment to any common standards or platforms. Data has been siloed in a wide range of legacy systems, some of which are difficult or impossible to support due to a lack of resources, documentation, or skill sets.

The lack of data centralization and data integration limits the value of the data and creates significant costs to maintain the databases and legacy systems. A few years ago, the AFMS decided that it had to modernize its data infrastructure to centralize and integrate its data.

- **Information deliverables**—Users of AFMS data sometimes feel they lack access to important data or must spend a significant amount of time finding, acquiring, improving, and personally integrating data to build the information artifacts they need. Not all information is delivered in a user-friendly way, it may take too long to access, and it is often designed just for a single-solution purpose. In addition, there is a lack of capacity (both technological and resource capacity) to support what customers want or need. As a result, many key consumers must be turned away to seek out other solutions.

- **Economies of scale**—Current tools lack the capabilities that are needed to quickly answer queries and create presentation-quality deliverables. The complexity of data has long and steep learning curves. The variety of technologies employed in data integration prevents economies of scale and hinders the development and institutionalization of standards and best practices.

- **Lag times from idea conception to realization**—Currently the total lag time from an idea for an analytical undertaking until a final output can take three to five years, sometimes longer. This long lag occurs as data has to be aggregated, resources are allocated, development and testing occur, information assurance (the government's term for data security) is performed, and implementation takes place.

- **Creating a skilled workforce**—A key part of modernization is growing a next-generation workforce that has the technical skills to use the data that is available.
- **Strategic alignment**—The OCIO faces the challenge of ensuring that all vendors, contractors, and key AF resources are aligned to a consistent, clear, and widely communicated IM/IT strategy and are empowered to succeed in the missions the AFMS has tasked them with.

Analytics at the AFMS

As with many large organizations, the ability to leverage data from disparate systems to provide usable information for operational and analytical reporting is an ever-growing challenge. Along with the general data explosion that has occurred, the healthcare industry has the unique challenge of standardizing data across health systems to enhance "point-of- care" delivery and use data for research analytics.

For over a decade, the AFMS has provided business intelligence (BI) services via two separate offices: the Health Informatics Division (HID) and the Health Informatics Suite (HIS). By providing registries and action lists, the HID enables clinicians to manage their complex and chronic patient population for effective disease management. The HIS develops solutions to assist the Healthcare Integrator (HCI) with initiatives such as provider schedule management, cost of care, and business planning.

The Health Service Data Warehouse Project

The Health Service Data Warehouse Project (HSDW) was driven by the challenges and pain points described previously. These challenges dictated the need for this data warehouse project to follow the approach of data infrastructure first and information delivery second, which focuses data integration resources on modernizing the HID's data acquisition and processing functions.

This "urban renewal" effort enhanced the HID's capabilities with the implementation of best-of-breed data integration software and more robust data architecture and infrastructure designed to scale for future growth and needs.

Previous attempts to modernize the AFMS's information assets focused heavily on consolidation and virtualization without focusing on data integration. While the colocation of database assets is important, it's equally (if not more) important to devote the time and resources to truly integrate the data. Without focused modeling and integration of the data, the organization will simply have all of its redundant information assets in one place without realizing economies of scale and the benefits of having one source of information.

The most critical factors for this project were

- **Data acquisition and transformation**—Previous data-acquisition processes used disparate technologies, were often antiquated, performed poorly, and were undocumented. Key knowledge workers to support the code have departed the HID, and the mix of technology skills required to sustain the current operations is too varied, increasing resource cost and hindering the creation of common technology standards. The solution involves transformation of the enterprise's data-acquisition processes to a centralized, completely integrated data warehouse.

- **Management**—Historically, business rules, metadata, and system documentation have not been centrally managed. HID capacity issues and varied technology have retarded the documentation process. Change management has frequently not been formalized, creating a moving target when process remediation projects are undertaken. And formal service level agreements for batch windows, performance, and system/data availability have not existed. In creating the HSDW, managerial processes have been revised to address each of these issues.

- **Information delivery**—Delivery of information has consisted largely of relatively static "push reports" or creations from analysts derived from hands-on data scripting and SQL Queries. Self-service BI has not existed. Along with the HSDW, the AFMS is creating specific data marts for various purposes (like the patient-centered medical home) and has created a portfolio of dashboards.

Implementation

A key to the success of the HSDW was having a champion in the organization who articulated a vision for BI and evangelized that vision. It shows the importance of having strong leadership support.

The HSDW implementation process took 12 months and consisted of the following steps:

- Requirements
- Design
- Development
- Test and Configuration
- Deployment
- Sustainment

Any data warehousing or BI project requires the right talent throughout the project life cycle. The key role facilitating the disparate groups involved in implementation is the analysts who bridge the invested parties: business, clinical, and technical. The many roles involved in the implementation process include a DW/BI lead; a work streams and requirements coordinator; lead and senior information architects; data modelers and analysts; SAS BI architects and developers; an SAS Center of Excellence lead and administrator; code and GUI developers; extract, transform, and load (ETL) architects, admins, and developers; and training, metadata documentation, and subject matter experts.

Results and Benefits

From a technology perspective, the architecture that has been developed is flexible enough to support both simple queries and complex analytics. The data now available can be accessed in near real-time. Users can analyze summary data and granular details.

The HSDW acquires, integrates, and stores the data once so that they can be repurposed. The logical components of a mature data architecture that support enterprise data warehousing and business intelligence include

- **Integrated HSDW**—Data are modeled and related according to business process and workflow.

- **Data marts**—Data are contextualized and accessible in a user-friendly form.

- **Operational data store (ODS)**—Data are persisted in near real-time for operational needs. This has the added benefit of streaming, which limits periodic batch extractions of significant size.

- **Data presentation/reporting**—There is a first tier of canned reports, charts, and tables to support simple users with frequent and recurring artifacts (SAS Enterprise Business Intelligence (EBI)).

- **Advanced analytics**—This includes a multitier BI tool capable of servicing complex queries and ad hoc data exploration and analysis (SAS EBI).

In addition to these technical benefits, having integrated data will enable clinical improvement, increase the satisfaction of users, and lower costs by reducing manual support time for disparate databases. It will also lead to greater efficiency through automation and will serve as a valuable asset for clinical research.

Lessons Learned

- In the AFMS, convincing all of the stakeholders to give up their information assets was a two-year process that eventually required an executive directive. An organization cannot underestimate the sensitivity associated with giving up control of data.

- Tool and technology selection becomes much easier when you realize that the talent acquisition to use the tool and/or technology is much more difficult.

- "Cyber warfare" will be a major technological trend to overcome.

- You can never do enough project planning, but too much planning is a threat to stakeholder perseverance, especially when they just gave up their information.

- The biggest mistake in planning occurred during the deployment of the HSDW's historical data load. This process exposed critical gaps within the existing architecture and infrastructure in the areas of role augmentation, storage capacity, and performance.

Next Steps

The focus of the HSDW on data acquisition, integration, and storage is critically important. However, to realize the full potential of data integration, the AFMS is focusing on data presentation, visualization, and delivery. The following actions are under way to deploy the "next level" BI to the AFMS:

- Using Informatics.
- Developing and delivering a platform for measuring clinical quality.
- Standing up an analyst-friendly SAS capability that allows analysts to focus on analysis rather than coding or data acquisition/integration.
- Developing a baseline for analyzing and delivering meaningful clinical research on de-identified data sets. Also, developing plans to support public/private access to data for research purposes.
- Providing intuitive user-friendly access to providers and managers to measure and improve their own performance.
- Creating reusable, repeatable processes and best practices around BI, and creating a BI Center of Excellence for sharing and developing AFMS-focused methodologies.
- Developing infrastructure and adopting new technologies to meet growing data, data warehouse, and BI needs.
- Integrating additional data into the HSDW.

22

Developing Enterprise Analytics at HealthEast Care System

Thomas Davenport

HealthEast Care System, an integrated provider network based in St. Paul, Minnesota, is the largest provider of healthcare services to the eastern metropolitan area of the Twin Cities. Consisting of three short-term, general acute care hospitals and one long-term acute care facility, it was created in a 1986 merger of several faith-based hospitals and home care organizations. In 2011 HealthEast had 7,000 employees and 1,400 physicians on staff.

In 2005, HealthEast embarked on a multiyear plan to become the benchmark for quality care in the Twin Cities. "The HealthEast Quality Journey," as the institution referred to its plan, was focused on making improvements in a variety of industry-standard clinical quality metrics, as well as internal metrics of process, operational, and workforce excellence. The HealthEast Quality Institute (directed by Dr. Craig Svendsen, Vice President and Chief Medical Quality Officer) was responsible for establishing goals and metrics. The institution's Informatics Department (directed by Dr. Brian Patty, Vice President and Chief Medical Informatics Officer) worked on incorporating improvements into everyday practice through the use of clinical information systems. The Medical Executive Committee addressed the topic of physician engagement with quality measures and care processes.

By 2010 and 2011 these steps had begun to result in substantial quality improvements. On almost all metrics, HealthEast had shown distinct improvement, and the provider led others in the market on key quality and patient satisfaction criteria. HealthEast focused particular attention on specific medical problems, such as ventilator-associated pneumonia (VAP). After implementing a set of process metrics and related order sets (the "VAP bundle"), VAP incidence improved dramatically. There were no incidents of VAP in 2010 in any HealthEast hospital.

In 2010, Thomson Reuters ranked HealthEast one of the top ten U.S. health systems based on a collection of clinical performance and patient care metrics. McKesson, a provider of information systems to HealthEast, gave the organization one of only two Distinguished Achievement Awards in 2010. According to the text of the award:

> Two years ago, HealthEast created a centralized command center using electronic tracking boards to help monitor patient flow in real-time 24/7 throughout the HealthEast hospitals. As a result, in less than a year, patient waiting times dropped in the emergency departments, patient satisfaction scores jumped 36 percent, and ambulance diversion hours decreased 63 percent.[1]

The HealthEast leadership was proud of its quality improvement achievements, but felt that there were additional areas to address if the organization were to continue its upward trajectory in care quality and patient satisfaction. One key area to address was enterprise analytics—the analysis and reporting of data across the entire enterprise, with a focus on prediction and not just reporting. It would also be increasingly important in the near future to integrate clinical, operational, and financial information. The pressure on U.S. providers to become accountable care organizations (ACOs) meant that clinical decision support and financial decision support would both need to influence patient care decisions. These capabilities existed independently at HealthEast and were difficult to integrate. Therefore, executives at HealthEast had been discussing the need to create an enterprise analytics capability.

Assessing and Integrating Enterprise Analytics Capabilities

The Informatics Department at HealthEast had begun to assess the organization's analytical capabilities as early as 2008. The department's leader, Dr. Brian Patty, believed that analytics were critical to HealthEast's continuing quality journey. He asked Skip Valusek, an industrial and systems engineer with considerable experience in process improvement and analytics, to assess analytical capabilities across the organization. Valusek conducted a survey of IT and managerial employees, and found that on a five-point scale of analytical capabilities, most respondents thought HealthEast was in the middle at Stage 3.[2] The surveyed group felt the organization had strengths in management sponsorship for analytics, analytical culture, and the analytical skills of HealthEast staff. The greatest weaknesses were judged to be in the data and information technology support for analytics. Valusek conducted another survey in 2010 and found similar results.

Patty and Valusek presented the results of the surveys at a regular monthly meeting of HealthEast's senior management team in early 2010. There seemed to be widespread agreement that the issue was important. Comments at the meeting included, "Enterprise analytics should be a component of our strategy," and "This is critical." In terms of implementation, someone pointed out that, "This requires a vision and steps."

The ownership of the enterprise analytics issue, however, was not firmly established at the meeting. Dr. Patty had initiated the discussion, but his organization was busy finishing the implementation of an electronic health record (EHR) for HealthEast. After the management meeting, Patty met informally with several senior executives who might have some interest in owning and managing enterprise analytics. None seemed to want to own the function. Patty concluded that it should be housed within his Informatics Department but that the establishment of an enterprisewide analytics organization would have to wait until the EHR had been fully implemented.

Designing the Enterprise Analytics Organization

By mid-2011, Dr. Patty felt the time was right to design and implement the new analytics organization. The EHR project was nearing completion, and in another executive session in July 2011 the executive team reiterated its support for enterprise analytics. However, the climate was somewhat less receptive for creating a new organization. Because of continuing pressure on reimbursements and the need for greater efficiencies in its care of patients, HealthEast needed to cut $50 million from its 2012 budget. A substantial number of new hires in analytics would be difficult to justify.

Therefore, Dr. Patty planned that most of the analytics staff would transfer into the department from other parts of the organization. This was feasible given that there were pockets of analytical expertise all around the organization. He envisioned three teams within the analytics organization:

- **Enterprise data team**—This team would focus on development and maintenance of a new enterprise data warehouse and data sourcing activities to yield "one version of the truth." It would include database administrators, ETL (extract, transform, and load) staff, and data architects. Some finance people who built the mature financial warehouse would transfer into this group. The enterprise data warehouse would eventually have data coming not only from the HealthEast hospitals, but also from clinics, home care, external EHRs, and physicians' offices.

- **Analytics team**—This team would use the data from the warehouse to do their analyses. Some of the people for it, including its proposed leader, Skip Valusek, would come from the informatics department. Others would come from existing groups within the organization where some analytics were already being performed. The Analytics team would not only generate analytics, but would also help with interpretation and process improvement based on the results of analyses.

- **Reporting team**—This team would develop a strong and largely automated reporting infrastructure, ending the current fragmented reporting and manual "data cobbling" practices.

Most of the personnel for this team would come from the Quality Institute and the informatics department. HealthEast used a tool called MIDAS+ for much of its quality reporting, which worked well except for the fact that there were often multiple sources of truth, even for Center for Medicare and Medicaid Services (CMS) reporting. Sometimes reports were also drawn from the wrong fields in the EHR. Automating these reporting processes would be critical for consistency and efficiency of data reporting.

Team members would have their primary affiliations to informatics, but would have "dotted line" reporting relationships to the departments they primarily supported, such as finance and quality.

Patty was anxious to get approval of the new organization and move forward with a higher level of analytical activity. He was particularly focused on connecting data across the continuum of care—sourcing, integrating, and analyzing data across the continuum of care. In terms of predictive analytics, he wanted to focus particularly on predictive models of readmission, working with HealthEast's case management function. And in terms of the organization's movement toward an ACO, he wanted to be able to report on clinical, financial, and operational metrics across the continuum of care for individual patients.

Notes

1. "HealthEast Honored for Improving Patient Care with Information Technology," press release, August 17, 2010, available at http://www.healtheast.org/press-releases/1107-healtheast-honored-for-improving-patient-care-with-information-technology.html.

2. HealthEast used a model derived from Thomas H. Davenport and Jeanne Harris, *Competing on Analytics: The New Science of Winning* (Harvard Business Press, 2007).

23

Aetna

Kyle Cheek

Healthcare payer organizations are undergoing a fundamental change to their traditional business model. Largely in response to cost pressures, payers are evolving from their traditional role as transaction-focused claims processing organizations to producers of analytically derived, information-centric consumer products. Payers have a long tradition of reliance on analytics tracing to their actuarial roots, but the recent transformation transcends the use of actuarial analytics to set rates and reserves. Rather, the move toward greater exploitation of value-laden information is centered on a deeper understanding of healthcare consumer behavior with the intent to influence behaviors and ultimately drive meaningful reductions in healthcare costs. A critical driver behind the heightened expectations for payers to provide information-product leadership is the simple fact that they are the primary consolidator of the largest volumes of the resource most critical to that enterprise—healthcare data.

While efforts to effectively harness the value latent in their information assets have met with mixed results, a handful of payer organizations have emerged as leaders in their analytic endeavors. Aetna is a leading example among the few payer organizations that can fairly be considered bellwethers in that space. Combining a deep analytical competency with mature data management practices, Aetna has positioned itself as a leader among healthcare payers and is able to leverage its analytical maturity for true competitive advantage. The origins of that analytical maturity in Aetna are unique, too, in that they have largely been developed organically. The following brief survey of

Aetna's analytics experience describes its origins, current organizational structure, and lessons for other payers that aspire to greater analytical maturity.[1]

History

Aetna's industry-leading analytics organization, Aetna Integrated Informatics, is owed largely to the acquisition of a deep analytical competency in another organization. Specifically, Aetna's current analytics organization traces to its acquisition of U.S. Healthcare, Inc. in 1996—an acquisition that included a subsidiary, U.S. Quality Algorithms (USQA). USQA was established by U.S. Healthcare in 1990 to provide insight into the billing and clinical practices of its contracting doctors and hospitals and identify opportunities to control medical costs.[2] Thus, with its acquisition, USQA provided to Aetna a mature healthcare analytics capability as well as a mature analytical staffing model. In addition, USQA provided an established data warehousing practice that provided infrastructure both for the acquired competency and on which Aetna's future analytical practice could be expanded.

Aetna's commitment to develop an analytical competency (including the competency acquired with USQA) coupled with a data-driven executive culture combined to position Aetna for a deep reliance on analytics. Oral histories of Aetna's informatics organizations commonly refer to the primacy of data-driven and metric-centric decision making as a key enabler of the success that has been realized through its unique set of capabilities. So deeply rooted is Aetna's reliance on this analytical capability today that it is generally viewed as a competitive differentiator and a "showcase competency."

Organization

Since its acquisition, the USQA analytics operation has evolved organizationally into today's Aetna Integrated Informatics. The responsibilities owned by Aetna Integrated Informatics have grown from the provider evaluation analytics that were the genesis of USQA and now provide analytic support across the enterprise. Reflecting its responsibility for analytic support across functions, Aetna Integrated Informatics is situated within Aetna's Innovation, Technology, and Services reporting structure and, reflecting Aetna's positioning of analytics as a critical component in its efforts to improve health outcomes and reduce medical costs, it also shares a dotted line reporting relationship to Aetna's chief medical officer.

Aetna Integrated Informatics' expanded scope of responsibilities includes five primary services for its internal and external constituencies:

- **Provider analytics**—This is comprised of outcome and cost analytics that are used to identify opportunities for outcome and cost improvements among physicians and hospitals.

- **Plan-sponsor reporting**—This constitutes the regular reporting on cost and utilization trends to covered groups (i.e., employer groups).

- **Program evaluation and research**—This consists of analyses that are routinely performed to assess the ongoing effectiveness of care management programs.

- **Custom informatics**—This is comprised of special projects.

- **Data warehousing**—This category refers to business ownership of Aetna's data warehouse.

Other core business functions that are aligned with the Integrated Informatics functions, especially through the enterprise data warehouse, are actuarial, underwriting, and marketing. Aetna's antifraud analytics are maintained apart from the analytical operations in Integrated Informatics.

Two critical features are apparent in the organizational alignment of Aetna's analytical operations. The first is that the core analytical competency (Aetna Integrated Informatics) is situated as a service organization, positioning it to provide broad support across the organization. Rather than virtualizing the organization, critical analytical functions have been consolidated into a single business unit—allowing greater coordination of analytical priorities under common leadership. Second is its business ownership of the data warehouse asset, which provides fundamental support for Aetna Integrated Informatics and also provides a collaboration point around which other enterprise analytic needs can be coordinated with the central competency.

Analytics Maturity Model

The Davenport-Harris Analytics Maturity Model articulates five levels of analytical competency to describe organizations' capabilities in this domain. From highest to lowest, the levels of analytical capability are

- Stage 5—Analytical Competitors
- Stage 4—Analytical Companies
- Stage 3—Analytical Aspirations
- Stage 2—Localized Analytics
- Stage 1—Analytically Impaired

Aetna clearly meets the criteria for Stage 4 competency and satisfies the Stage 5 criteria in several areas.[3] Perhaps most importantly, Aetna has identified those analytics that are most critical to competitive advantage in a comprehensive analytical strategy. Additionally Aetna has consolidated responsibility organizationally for those priorities that provide a consolidated and aligned view of resource needs (human and technical) and adds the benefit of a largely consolidated enterprise analytical toolkit. Importantly, Aetna also has a strong competency in the critical data warehousing infrastructure and business ownership that is required to drive large-scale analytical efforts. The value of strong leadership around data management requirements by the analytics organization is apparent in Aetna's planned migration of its warehouse from a plan- and product-centric model to a member-centric model. That effort is positioned for greater success by virtue of the involvement of a critical stakeholder. Similarly, the analytics organization is positioned to better evolve its analytical agenda by virtue of its positioning vis-à-vis requirements for the critical data infrastructure.

Isolated examples do remain of analytics that are not fully consolidated into the centralized informatics operation, namely actuarial, underwriting, marketing, and antifraud analytics. However, most of those examples are aligned with the centralized analytics organization through its maintenance of data warehouse content requirements—a basis for alignment that ensures that those stand-alone analytic areas share a common understanding of analytic priorities and resource requirements.

Aetna's placement near the top of the maturity model makes it a clear leader among healthcare payer organizations, especially considering that most payer organizations are best classified at Stage 2 on the Davenport-Harris maturity model. Unlike Aetna, most payers have some localized capabilities but do not have a holistic analytics strategy. Most payers also lag Aetna's maturity in the data warehousing space and do not have an enterprise analytics toolkit. Given the context of most organizations in the payer space, Aetna's maturity is all the more apparent, as is its competitive differentiation with its deep analytical capabilities.

Bellwether Lessons

The most apparent lesson to be learned from Aetna's bellwether position among payer organizations may be more broadly applicable than the conspicuous differential in size and scale between it and other payers that, with few exceptions, operate on a smaller scale in more limited geographies. The first lesson for aspiring analytical payer organizations is the importance of identifying the strategic drivers that offer the most demonstrable value from analytical enhancement. For Aetna, that strategic clarity was to some extent encompassed by the acquisition of the analytical competency—that is, Aetna's acquired competency already had a focus at the time of acquisition. This established focus in turn provided a foundational basis for Aetna to expand its reliance on analytics to other business functions.

Another important consideration that emerges from Aetna's case study is the importance of the relationship of the analytics organization to the underlying data infrastructure on which analytics are dependent. For Aetna that is seen specifically in the traditional role its Integrated Informatics organization has played as business owner of the data warehouse. By virtue of its formalized stakeholder role,

Integrated Informatics can directly influence decisions that impact the data resources on which it relies. Integrated Informatics' role as business owner of the enterprise data warehouse also allows a holistic view of evolving analytical priorities and needs, such as the migration of the warehouse to a member-centric view to drive more patient-centric outcome and cost containment analyses.

The importance of organizational placement is also apparent in the Aetna case study. While Aetna's organizational model surely will not translate directly to every payer, it does underscore the importance of situating a comprehensive analytical competency such that it is able to serve a broad-based constituency within the organization. In the Aetna example this critical consideration around organizational placement intersects with the extent to which the stakeholder role has been formalized around the management of data assets. While Integrated Informatics is positioned to provide analytical support to a broad constituency, it also provides coordination beyond its direct customers by owning responsibility for the data assets on which the entire organization relies.

Finally, the Aetna experience speaks to the importance of developing an internal analytics competency for payer organizations. Aetna capitalized on a unique opportunity that presented itself from an acquisition. But the more important lesson may be found in Aetna's commitment to internalize the competency it acquired and integrate it across the organization to drive broad-based analytical value. By consolidating its analytical competency in one organization Aetna promotes alignment of priorities and efficiencies in the management of the underlying infrastructure on which analytics are dependent.

Payer organizations that aspire to analytical maturity would be well-served to consider these critical elements behind Aetna's bellwether use of analytics for competitive advantage.

Notes

1. Except as noted otherwise, this chapter is based on a November 23, 2010, telephone interview with Brian Kelly, MD, Head of Informatics and Strategic Alignment at Aetna, and Kathe Fox, PhD, Head of Consultative Informatics at Aetna.

2. "U.S. Healthcare, Inc., History," http://www.fundinguniverse.com/company-histories/US-HEALTHCARE-INC-Company-History.html.

3. International Institute for Analytics (IIA) official benchmarking study not conducted; competency score and conclusion derived from author.

24

Employee Health and Benefits Management at EMC: An Information Driven Model for Engaged and Accountable Care

David Dimond and Robert Morison

EMC Corporation is the world's leading developer and provider of information infrastructure technology and solutions that enable organizations of all sizes to transform the way they create value from their information. Headquartered in Hopkinton, Massachusetts, EMC has annual revenue of $16.5B and 45,000 employees around the world.

EMC's Driving Partnership in Health program has transformed how the company promotes workforce health, how employees consume healthcare services, and how much the company pays for those services.

- Since 2004, EMC's healthcare costs have been well below the national trend line, amounting to cost avoidance to date of $223M (see Figure 24.1). The company achieved that dramatic cost containment while expanding services and without shifting cost to employees.
- More than 90% of EMC's 22,000 U.S. employees are registered on HealthLink, the company's personal health management portal, and 95% of them have taken a Health Risk Assessment (HRA). An astonishing 94% of covered spouses and domestic partners have done the same.

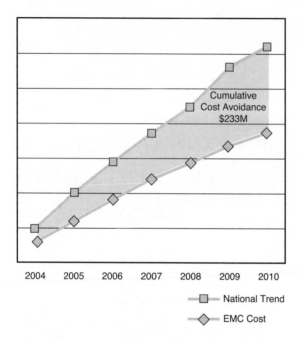

Figure 24.1 EMC healthcare costs and cost avoidance

- EMC was the first employer in the world to sponsor an electronic and automatically updated personal health record (PHR) for employees who choose to maintain one. More than 40% of employees are doing so on a regular basis.

- The company's data warehouse of employee health information enables the personalization of employee experience as a health services consumer, the aggregate analysis of healthcare consumption and cost trends, and the identification of opportunities to lower cost by improving workforce health.

- Its efforts have earned EMC recognition and awards from, among others, the Massachusetts Health Council, New England Employee Benefits Council, Mass Technology Leadership Council, and a UnitedHealthcare Apex Award for innovation in the healthcare experience of employees.

- The company does not keep its success a secret, but rather presents regularly at industry conferences and hosts visits from healthcare providers, other employers, and others interested in learning from EMC's experience. By working closely with the partners in its employee healthcare ecosystem, as well as sharing its experience, the company strives to shape the evolving healthcare landscape.

Vision and Lessons Learned

EMC envisions a world where employers can engage patients and providers, enable health awareness and literacy, influence health and lifestyle behaviors, and drive adoption of patient-centric technologies. The results should include healthier and more productive employees, containment of healthcare costs, and societal change around the delivery and consumption of healthcare services. The right marriage of informed and responsible employees with technology-enabled services should improve communication and increase patient involvement in the healthcare process; improve efficiency by eliminating both time delays and duplication of records, tests, and procedures; and improve patient safety by reducing manual error and enabling more sophisticated vetting of diagnoses and prescriptions.

Why invest actively in a healthy workforce? Because it means greater productivity and lower total cost through reduced claims, reduced absenteeism, enhanced engagement while at work ("presentism"), and lower turnover. As EMC has demonstrated, an employer can bend the healthcare cost curve favorably by influencing utilization, by steering informed employees to make healthy decisions and to choose high-quality and cost-effective options for care. We've mapped EMC's experience in this information-driven program into a preliminary "needscape" as shown in Figure 24.2.

Through the course of developing this program and the culture that surrounds it, EMC has discovered the following:

- **Employee engagement**—There are relatively simple descriptive analytics that represent the status of a given employee or a group of employees in a panel. With the trend toward accountable care and the need to activate, engage, and perpetually improve the health of chronically ill or at-risk patient panels, trends in this high-value data are featured in management dashboards.

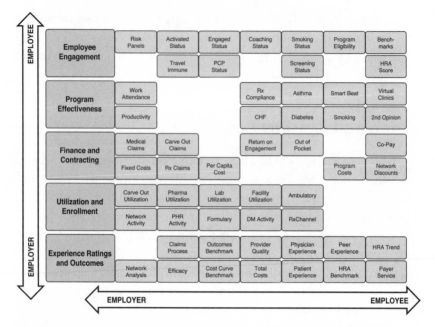

Figure 24.2 Preliminary "needscape" based on EMC's information-driven program experience

- **Program effectiveness**—EMC participates in innovative programs to promote wellness and to manage chronic conditions, and shares the results as a means of analytically informing and motivating employees to participate. A striking example is how a panel of hypertensive EMC employees served in a clinical trial that was the genesis for a program founded with Partners Healthcare's Center for Connected Health called SmartBeat.

- **Finance and contracts**—These are the key analytical competencies one would expect in a plan sponsor working with its broker to make decisions related to performance improvement, cost containment, and contracting. As a "big data" company, EMC has a unique opportunity to begin working with analytics partners, with de-identified data, in the areas of predictive analytics and even causal modeling. These types of studies will guide critical decisions related to EMC's plan tiering and serve as precursors to accountable care management methods.

- **Utilization and enrollment**—The utilization of services, programs, and clinical resources is tracked and trended over time and serves as a critical feedback loop to support decision making and discovery.

- **Experience rating and outcomes**—These kinds of analytics are common in the music and travel industries, where consumer experience, quality, cost, and productivity (efficacy) rule—and drive adoption and (sometimes disruptive) innovation. As variation in practice decreases with the precision of treatment, people will make decisions about their care based more on such consumer-oriented inputs.

EMC's level of maturity, when aligned with the Davenport, Harris, and Morison Analytics at Work framework, is summarized in Figure 24.3, which shows the typical types of questions EMC needs to address.

	Past	**Present**	**Future**
Information	What is the historical claims cost for Lower Back Pain (LBP)? (reporting)	What is the cost this month for all Massachusetts based employees? (alerts)	What is the forecasted trend in with Massachusetts employees with a BMI>30? (extrapolation)
Insight	What is the statistical relationship of BMI to LBP claims in Massachusetts (modeling)	Develop and launch Pedometer activity tracking "Steps" Competition for Employees (recommendation)	Identify and quantify drivers of LBP and intervention response (prediction, optimization simulation, hypothesis free discovery)

Figure 24.3 Questions EMC needs to address

Chronology

EMC's approach may seem enlightened and sensible today, especially after the intensive scrutiny of healthcare delivery in the process of enacting national health reform legislation. However, when Delia Vetter, senior director of benefits, and Jack Mollen, EVP of human resources, began exploring new approaches to employee healthcare almost a decade ago, their peers were naysayers. When Jack keynoted a Conference Board meeting in 2003 and talked about EMC's vision of consumerism in employee healthcare, the feedback said, "It will never work," and "Employees will revolt over privacy issues."

In 2002, Delia, Jack, and EMC's senior management recognized that double-digit annual increases in employee health costs were untenable for both the company and its employees (costs were projected to double in five years), and that past approaches to cost containment hadn't accomplished much. They needed to do something different to contain cost without simply shifting cost to employees (as most companies have chosen to do).

EMC is self-insured, and it worked through its healthcare plans to get the best available rates, so there was only so much it could do to influence unit costs. Therefore it had to focus on the utilization element of the cost equation: influence utilization by encouraging workforce health, informing employees about healthcare choices and the true cost of services, and focusing on common, chronic, and potentially costly conditions such as hypertension.

The problem was that the company lacked data and analytics about consumption of healthcare services—data was fragmented across 38 health plans across the United States. And no large company had experience to share implementing the approach they had in mind. To understand the challenges and feasibility of new approaches, Delia got involved in healthcare coalitions, business roundtables, and health plan and provider advisory boards—anyplace she could meet and learn from progressive practitioners. In the process, she also found credible experts to invite to EMC to talk with its senior management team.

Meanwhile, her team gathered data from the health plans to draw an initial picture of healthcare consumption at EMC and identify some initial opportunities to improve employee health while containing cost.

The initial strategy was developed with the help of EMC's insurance broker and advisor, Willis, in 2002. By 2003, they had worked with WebMD to launch the first (mainly informational) version of the HealthLink portal. They also worked with Ingenix to build a data warehouse consolidating claims information from the 38 plans. In 2004, they introduced a more interactive portal, including the basics of the personal health record—services, prescriptions, diagnoses, costs. Delia recalls: "At first we didn't realize we were assembling a comprehensive and portable PHR. We were just trying to get useful information about consumption and true cost into the hands of employees. Then we realized that we were breaking new and important ground."

As the data warehouse added information, it enabled EMC to analyze aggregate consumption and cost patterns and find specific opportunities for promoting preventive care and employee health. It also provided the detailed information to enable interested employees to understand true costs and become more effective consumers. Few employees were aware of the cost of services beyond their deductibles and co-pays.

By 2005, HealthLink and the new approach had become institutionalized, the benefits group was working to an annual roadmap of coordinated initiatives, and the program began to receive external recognition. In 2007, HealthLink was opened to family members. In 2008, the PHR was significantly improved to import lab results, enable optional provider access, and enable portability. The Health Risk Assessment was introduced in 2008. Since then, the portal has continued to add capability, with an emphasis on information exchange and integration with providers, as well as functions for specific programs such as remote patient monitoring.

What's happening today or on the drawing board? Radiology/imaging results, additional lab tests, provider notes, remote monitoring, and biometric readings are being incorporated into the PHR. Virtual clinics staffed by nurse practitioners are being established at major locations, both to save on doctor visits and to encourage employees to get help when they need it. Remote monitoring is expanding to include diabetes patients. The list of available health programs, clinical studies, and online tutorials continues to grow. HealthLink is going mobile with access and applications on personal devices. EMC is working with providers on tele-health initiatives and anticipating the eventual integration of the PHR with industry standard electronic health records (as the latter are defined). With the help of its own state-of-the-art products, EMC is enhancing the security of the PHR and all of its healthcare management infrastructure and interfaces.

EMC is planning to roll out an appropriately revised program to employees in Canada and is working on a vision and approach for more actively promoting employee health worldwide. The company continues to participate in health transformation initiatives in Massachusetts and nationally.

Along this journey, there were two keys to success:

- EMC leverages a wide range of partner organizations: Willis, WebMD, Ingenix, major healthcare providers, major health plans, pharmacy managers, medical research institutions, specialized diagnostic and care providers, and business and industry associations. EMC has built an ecosystem of organizations (some in competition with one another) mutually committed to innovate and cooperate in support of EMC's strategy.

- EMC has stayed true to that strategy and its long-term objectives. The list of initiatives may be lengthy, but they fit together. Nothing is rolled out in isolation. Everything is evaluated and implemented in the context of leveraging technology to improve employee health.

Employee Experience

EMC's goals for its employees center around activation and engagement. That includes taking advantage of available resources, taking responsibility for being an effective consumer of healthcare services, and learning about one's state of health and how to live a healthy lifestyle.

EMC employees have access to a wide range of self-service tools and resources, support and encouragement in using them, specific incentives, and controls. The introduction to these resources—and to the health management partnership that EMC wants to forge with employees—is a nontraditional health benefits booklet.

The booklet is "nontraditional" because, while it includes the standard enrollment timetable and overview of plans, premium costs, and special programs, the details and comparisons are maintained and accessed online. The booklet instead focuses on EMC's health philosophy, Driving Partnership in Health: "We believe that being good consumers of healthcare means focusing on healthy lifestyles, patient safety, quality of care, and understanding the impact we, as individuals, have on healthcare costs...choice and responsibility are the foundation."

The booklet also discusses healthcare reform: how the Patient Protection and Affordable Care Act changes responsibilities of both employers and individuals. It lays out the timeline of healthcare reform provisions, highlights those that EMC has already met (in part due to earlier reform in Massachusetts), discusses implications for employees, and generally educates them on the changing healthcare environment.

The portal to most resources is HealthLink, an online personal health management site for employees and their family members, secure and available 24/7. Central to the health management process are

- **Personal health record**—Enables employees and adult family members to review clinical data and discuss them with caregivers, in the process avoiding duplicative tests and procedures, minimizing untoward medication interactions and side effects, and generally participating more actively in their treatment. The PHR is automatically populated (via the data warehouse) with medical information including provider, date of service, diagnosis, lab results, prescriptions, and cost of service (both actual and out-of-pocket). The PHR can receive remote patient monitoring information, and it is downloadable and portable.

- **Health risk assessment**—Survey provides participants with direct feedback on a series of risk factors, including stress, nutrition, weight, blood pressure, blood sugar, cholesterol, exercise, and alcohol, tobacco, and substance use. The HRA yields a total score (on a 100 point scale), a simple means of focusing attention and a benchmark as people try to raise their scores year over year. Linked to the HRA are guidance on behavior change and lowering health risks, as well as the additional costs associated with specific risk factors. It's not uncommon to hear EMC employees say "What's your number?" to others who openly discuss having opted into health improvement programs.

HealthLink provides access to a vast array of medical news, information, and recommendations maintained by WebMD—and automatically suggested based on the individual's health profile. HealthLink also archives EMC's monthly health seminars. And it provides important functionality such as

- Hospital and physician selection based on condition, procedure, and location, as well as "best doctors" ratings
- Health alerts, including about potential drug interactions
- Automated way to seek a second opinion
- Healthcare cost tracker broken down by category (e.g., physician, pharmacy, lab) and highlighting both employee and EMC costs

EMC employees and families have access, at no added cost, to a variety of specific health management programs and facilities, for example:

- **DASH for Health**—A dietary program for reducing blood pressure and attaining other health benefits.
- **SmartBeat**—Remote patient monitoring of blood pressure. EMC participated in the initial clinical trials and helped demonstrate the value of self-management in reducing high blood pressure.
- **LiveHealthy**—A customized program, including individual coaching and support, for those at increased risk for or currently living with a serious or chronic condition such as asthma, diabetes, or obesity.
- **Quit for Life**—Smoking cessation program.

EMC sponsors a variety of onsite health management seminars, often supplemented with webinars and support groups. And it offers state-of-the-art fitness facilities at major locations.

The direct incentive for employees and their family members to engage in their health management is a reduction of about 10% in monthly premiums. To qualify for the lower payment in 2011, employees and their spouses or domestic partners must only complete the HRA. To qualify next year, they must also complete biometric screening (that automatically populates parts of the HRA) and participate in the "Choose One and Commit" program. The latter simply asks the employee to do something healthy (e.g., participate in a community walk or softball league) or make use of a specific resource (e.g., hospital advisor, mammogram screening, prostate cancer screening, or just maintaining one's PHR). The "One" activity can be chosen from two dozen options that include the formal programs such as DASH.

Indirect incentives include the encouragement of colleagues and family members. When a spouse or partner completes the HRA and compares results with the employee, it can lead to strong mutual commitment to raise the scores.

Some employees have hesitations and concerns about the health benefits program. Some are simply uncomfortable with their employer's activist role in what has traditionally been the private matter of maintaining health. This concern has been addressed with a great deal of education, including individual discussions when needed and consistent, persistent messaging about the purpose and terms of the employee-employer partnership.

Others have particular concerns regarding the security and protection of their health information—no surprise since EMC is in the information management business and has experts on the payroll. This is addressed by explaining the safeguards in place and keeping them state-of-the-art using EMC's own technologies.

Concerns are also mitigated through controls, and the employee is ultimately in control. Employees can opt out of the health benefits program (though already in Massachusetts and nationally in 2014 individuals are required to carry health insurance or pay a penalty). They can pay the full premium and not avail themselves of the HRA, PHR, or other resources. Or they can participate and exercise specific controls over, for example, whether to automatically import claims data to the PHR, whether and when to permit physician access to the PHR, whether to receive health alert messages, and, of course, security and access settings.

As employees gain insight into the costs of their healthcare—and recognize how EMC has maintained coverage and expanded programs without cost shifting—they buy in to the program. One of the surest signs that the partnership is working is that surveyed employees list EMC as their most trusted source of health information.

Partner Perspective

What's it take to partner with EMC in its program to manage employee health and benefits?

- **Commitment**—Not just to provide excellent standard services, but to work with EMC in meeting its objectives. To determine who was really on board, EMC asked all of its provider and plan partners to rebid for their contracts.
- **Collaboration**—All the partners, including competing providers and plans, must participate effectively in EMC's employee health ecosystem. To build this network, EMC convened a Partner Summit of dozens of organizations to share information, align with EMC's goals, and work together.

Examples of partner experiences are discussed in the following sections.

Advisor

Willis is a leading global insurance broker and consultancy with a history of innovative approaches to risk management. Bill Schlag and a Willis team have been working with EMC for a decade to shape and execute the strategy to engage employees, empower them with information and services, and outperform the healthcare cost trend. In addition to providing ongoing broker services with providers and plans, Willis has advised and participated at every stage of EMC's journey—from claims analysis to target common and costly conditions, to organizing employee health and claims data into a repository, to launching and improving the portal, to designing and maintaining the dashboard for executive reporting.

Provider

The director of clinical informatics at a major regional health-care system shared a provider perspective. The organization has an enviable record of quality care and is focused on increasing access to health services to improve prevention and reduce chronic conditions. Thus, its partnership with EMC is rooted in closely aligned goals and mutual ambition to improve. They recognize the limits of what can be done in a doctor's office (and that many people are too busy to schedule appointments), hence the value of bringing services to patients (e.g., onsite clinics, in-home monitoring) and influencing patient behaviors. Technology is helping with an attitudinal shift: "I can participate in my healthcare anytime." The HealthLink portal has a gateway to the provider's patient portal. And personal health records are transferrable (manually today, with a direct interface in the works).

Plan

For a major healthcare plan, the working relationship with EMC is "definitely not standard." In addition to the basic role of structuring coverages and participating in care management, the organization is challenged to break down barriers to sharing data (including claims and performance metrics), and to support a variety of special initiatives (from the DASH program to onsite immunizations for EMC employees traveling internationally). For example, a specific data exchange problem relates to populating the PHR with lab test results when some (e.g., pregnancy, HIV) are highly confidential and restricted. The onsite immunizations required streamlining a claims process.

For the plan, EMC is a highly respected reference account. The plan also appreciates how EMC is "rallying the provider community" around preventive care and patient participation. The plan has to find the resources to do new things with EMC, and it "can't work with everyone this way," but the opportunities to learn and innovate make the relationship very worthwhile.

Program

DASH for Health (Dietary Approaches to Stop Hypertension) is administered by the Boston University School of Public Health. The original trials of a dietary approach to lowering blood pressure were conducted in Boston and then around the country in the 1990s. Following the diet, which emphasizes fruits and vegetables, can be the equivalent of taking a standard daily dose of hypertension medication. It has also proven to lower cholesterol, improve bone density, and enable people to lose weight and generally feel better. In 2007, EMC employees participated in a research study demonstrating that for people with specific cardiovascular risk, adhering to the DASH diet can lower healthcare costs by an average of $800/year.

Through claims analysis, EMC had determined to focus on hypertension, so the DASH program was a natural fit. The program is administered mainly online, through HealthLink and the DASH site, though DASH staff also provide onsite orientations. Because success with such a diet can depend greatly on having supportive friends, a recent addition has been the opportunity for EMC employees to invite up to three friends to participate with them.

Platform

For eight years, WebMD has partnered with EMC to deploy and enhance the HealthLink portal. HealthLink is a privately branded version of the WebMD portal configured to support the HRA, PHR (portable to WebMD.com), and a variety of self-directed health and lifestyle improvement programs. The portal also incorporates WebMD's reference information on conditions and treatments, and important functions such as health alerts (e.g., about potential drug interactions).

Working with EMC provides the opportunity to continue to innovate, especially in "keeping the healthy healthy." Ongoing initiatives include more seamless access to better integrated data, improved self-service and self-reporting by patients, and better analytics to understand the needs of patient cohorts, evaluate the ROI of programs, and improve health management for the entire covered population. WebMD works with other progressive companies, but EMC is the leader in deploying PHRs, developing innovative solutions, and bringing its partners to the table.

Common Threads

Three common threads run through these partners' experience working with EMC:

- Aligned objectives, including a commitment to quality, cost-effective care, and preventive care
- Willingness to share both information and expertise
- Opportunity—and imperative—to innovate

The Partner Summit was an eye-opening experience for all involved. It brought together providers, plans, pharmacy managers, medical research organizations and their programs, technology providers, specialized clinics and other service providers, and even the providers of disability and life insurance. This range of players had never gotten in the same room, let alone for purposes of brainstorming about the future of employee healthcare and how to make it better. EMC's attitude was not, "Here's how we're going to operate," but rather, "We're here to learn from you—what can and should we collectively be doing additionally, differently, and better?" That stance got the players talking, looking at the data, and working together. Erstwhile competitors were co-opted to behave as co-implementers.

Executive Scorecard

Not all senior executives think of a healthy workforce and healthcare cost management as factors critical to business success. Given health and healthcare cost trends, they should, so ongoing communication and education are in order. Delia Vetter's group works with Willis to issue a quarterly scorecard for EMC's senior management. It includes

- Health plan costs—aggregate, per capita, and trends. These are also broken down by claim type (medical, pharmacy, behavioral health) and health plan.

- How EMC and its employees share those costs.
- Employee and covered member demographics.
- Top ten listings with costs of hospitals, diagnostic categories, pharmaceuticals prescribed, and therapeutic classes.
- Prescription drug breakouts—generic versus branded, retail versus mail order.
- HealthLink portal usage statistics and trends, including aggregate HRA results.

The scorecard also contains a narrative overall assessment plus brief updates on key initiatives, accomplishments, healthcare topics, and other observations. And it incorporates an illustration of EMC's three to five year healthcare strategy and mention of awards received as an indicator of influence in the marketplace.

The scorecard is four pages long and distributed both electronically and as a laminated document for the convenience of the several executives who like to carry and discuss it regularly. It informs and educates executives and helps align them around employee health strategy and initiatives. It also makes the value on ongoing investment—including in the technologies of healthcare management—very clear.

About the Data Warehouse

The EMC data warehouse, from their partner Ingenix, aggregates claims and enrollment data from each of EMC's vendor partners, including ADP (eligibility), Medco (Rx), the self-insured medical plans, a behavioral health plan, short and long-term disability insurer, and workers' compensation, plus HRA information from the Health-Link portal. The data are updated monthly.

Ingenix scrubs the data and loads them into a tool called Parallax i that EMC and Willis use to measure, analyze, and evaluate EMC's benefits programs (see Figure 24.4).

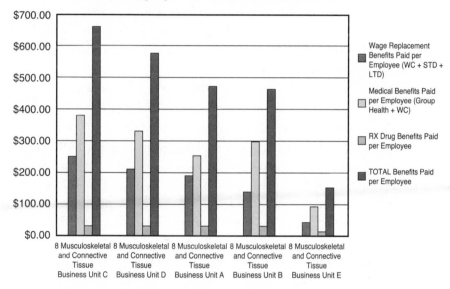

Figure 24.4 Average costs per employee by benefit type

The data are also grouped into episodes of treatment for analytic purposes. Data feeds go to HealthLink to populate the PHR and for the Health Alerts messaging program, to Medco for management of the prescription drug program and drug interaction analysis, and to EMC's disease management vendor.

EMC in the Larger Context

EMC's experience and accomplishments offer lessons both for other large corporations and for the healthcare system at large. How EMC uses information and technology to contain cost and promote employee health confirms some of today's major trends and previews the likely experience of others.

Data is the foundation; trust is the key. With good data, an organization can analyze the patterns of consumption, cost, and medical conditions, then focus on the problem areas. It can isolate cost drivers and manage costs. It can inform and steer employees to be mindful of quality, convenience, and cost—and to make healthier lifestyle choices. And it can continuously measure and analyze what actions are most effective in promoting workforce health and containing healthcare costs. However, the challenges of data acquisition and integration can be herculean. Providers and plans can be reluctant to share cost, price, and quality data, having long treated it as a proprietary asset. And they've sought the cover of privacy concerns.

Today, however, we have the means to protect information in the form of advanced and analytically driven adaptive authentication technology, which is part of the next generation PHR platform. EMC knows from the world of financial services that there is a strong correlation between consumer trust and online adoption, so folding in this technology, which is the de facto standard in online banking, just makes sense. A foundation of data, wrapped in employee trust, enables collaboration, information sharing, and discovery that capitalizes on the new transparency.

There are sufficient data to gain experience in the rapidly emerging model of Accountable Care. EMC has implemented many of the principles and practices associated with Accountable Care Organizations (ACOs):

- Analyze and address the needs of specific patient populations segmented by condition or risk factor.
- Focus on preventive and as-soon-as-needed care, thus reducing preventable procedures and hospital admissions and containing cost both short-term and long.

These approaches work for employers and providers, even when the latter are not structured and contracted as ACOs. They simply make sense as the means of simultaneously improving health and containing cost.

Employers should step up, particularly Health Delivery Organizations considering ACOs. Most large corporations and even non-profit hospitals have been too much on the sidelines of optimized healthcare delivery, offering employees standard (and often shrinking) benefits while absorbing regular cost increases. In the process, some have reneged on obligations to employees and retirees. EMC has demonstrated the value of purposefully participating in and influencing the healthcare delivery process, of being activist on behalf of both employees and shareholders. A key lesson here is that there are no quick fixes, technological or otherwise. Electronic health records (EHRs) and health information exchanges (HIEs) are not going to get all the data in order, and the organization that plays wait-and-see just sees its costs rise. The employer has got to commit and act locally with its employee data, but with an eye toward eventual interface or integration with EHRs and HIEs.

Healthcare is becoming patient-centric and employers will activate and engage their employees with health information centered "medical homes." Consumers are being empowered with information and choice. There's no fighting the trend, so employers—indeed, all the players in the healthcare ecosystem—must embrace it. For employers, that means partnering with employees to build their healthcare literacy and encourage them to make sensible choices. That also means using technology to understand the employee-consumer's experience, set new standards for information sharing and transparency, and provide services compatible with how employees live, work, and behave. More of these services are accessible or consumed in the workplace and in the home because that's both convenient for consumers and cost-effective for providers.

To compete on analytics, the stakeholders across the value chain may be challenged by the influx of analytics-enabled employers with engaged employees who will behave like savvy consumers in an era of choice and responsibility.

25

Commercial Analytics Relationships and Culture at Merck

Thomas H. Davenport

The Commercial Analytics and Decision Sciences group at Merck is responsible for assisting with advanced analytics for all of the U. S. primary care, hospital/specialty, and vaccine products at Merck. Its focus is sales and marketing analytics, including customer targeting, segmentation, sales force sizing, promotion response modeling, ROI assessments, and other related analyses. The group's primary mission is to help senior leaders at Merck make better business decisions with regard to multimillion dollar promotional and sales budgets. It has existed at Merck for more than 15 years.

The group consists of more than 25 full-time staff and several other external consultants. The leader of the group has a Ph.D. in Applied Research and Evaluation, and most staff members have advanced degrees in quantitative fields, including operations research, statistics, and economics. Most of the group's staff came to Merck from analytical roles in other companies spanning numerous industries, including consulting, large pharmaceutical firms, health services research, physician licensure, insurance, and consumer packaged goods. In addition, the group maintains close partnerships with a variety of external providers of data and analytical software and services.

The Commercial Analytics group has been involved in a variety of key decisions at Merck over the last several years. When Merck reengineered its commercial model for U.S. Sales, the group piloted the model before it was adopted, with test and control groups. Other work has quantified the impact and profitability of virtually all major investments in the physician and consumer channels. The group has also created tools for optimizing sales force sizes and structures along with multichannel programs.

Decision Maker Partnerships

The Commercial Analytics group maintains a close set of relationships with internal business decision makers. They have positive comments about the group's role. One executive, responsible for strategy execution, commented:

> A lot of times Commercial Analytics team members were my "thought partners" in implementing the new field organization. Working with them was a good way of thinking something through. We used them as sounding boards. They are very solid problem solvers, and play the role of an objective third party.

The same executive said that the Commercial Analytics team was more useful than an external resource that did similar types of work:

> Most of the other firms who did these new commercial models used an external consulting firm. We used them for some tasks, but we had our own algorithms developed by Commercial Analytics. They also found ways to optimize and test the pilots. It gave us a better result, as well as more internal buy-in.

The leader of a new business area that worked with Commercial Analytics also had positive comments about the value of the group's work and their credibility:

> Our business area is a pilot program. We want to show that it drives new revenue and provides better customer support. Commercial Analytics is measuring the impact of the pilot program. They set up a rigorous test-and-control approach.... Commercial Analytics is very familiar with the business. They ask what business questions you are trying to answer, and

then they identify how to measure them. They will analyze the data to see if they can answer the questions. Their level of objectivity is what you need to have; we need an independent source.... At times in the past, Commercial Analytics had to tell senior management that their project doesn't have good ROI. They are very credible when they do that. And if they say it works, there won't be any doubt about it.

A senior executive at Merck with global responsibilities emphasized the value of having Commercial Analytics involved in the entire decision process:

They should always be at the table when we are making an important decision. I remember when we were evaluating the returns on a major promotional campaign a while back. Commercial Analytics was at the table with us throughout the discussion, and would engage with us in debate. Then they would do analysis to answer key questions. Having them be part of the team is a real competitive advantage for us.

Reasons for the Group's Success

There are undoubtedly many reasons why the Commercial Analytics and Decision Sciences group at Merck has been effective. The members of the group certainly have a high level of analytical skills, for example. Another key factor, however, is clearly the culture and relationships orientation in the group. The leadership team of Commercial Analytics emphasizes the key value of the organization:

The umbrella over everything we do is a culture of motivating team members with the prize that we are here to help our clients make better decisions through the use of our analytic insights and tools. Our rules of engagement are to make your internal client understand that you are there to help them make a better business decision.

The cultural orientation begins with clarity about the mission and responsibility of the organization. The group leader notes:

We're always objective about our findings. In a way we are the "Switzerland" of marketing and sales at Merck, providing a neutral perspective on those decisions. We work for the shareholders.

The group leader gives an example of how the group's independence affects its work with internal clients:

A lot of times managers will hear that we can do ROI analysis on promotions. So they come to me and ask if we can help them. I say, "We can do that, but let me ask you a question first. We will find that your promotion was very effective, marginally effective, or ineffective. Can you tell me what actions you'll take in each of those cases?" We document their answers and how the analytics will tie to them.

The Commercial Analytics leadership team refuses to have the group engaged in a project if there is no clear relationship between an analysis and the decision to be made.

The group's strategy execution client confirms this approach:

Commercial Analytics always has the question of, "How's this going to add value to our business?" at the front of their minds. They aren't chasing stuff as an academic exercise; they do a good job of checking to make sure that what they are about to tackle has business value. They ask, "What are you trying to get at—maybe there is a better way to get there." I don't think it bothers anyone when they push back a little—they do it in a nice way.

One of the reasons that the group is able to work successfully with business decision makers is its emphasis on clear and nontechnical communication about its work. Many of the analyses it undertakes are technically complex, but the Commercial Analytics leadership team devotes considerable effort to translating them into straightforward business terms. The leader of the group describes this process:

We work hard at packaging our results in a way that is very intuitive and easy to digest for our business clients. If an analysis is not understandable to our client, then we failed to provide the appropriate graph, chart, or table. We do not avoid complex methods, but we make sure we can explain them. One of our passions is distilling very complex ideas into simple terms so that businesspeople can understand and apply them.

The executives at Merck interviewed confirmed that the communications approaches are succeeding. For example, the executive leading the new business area noted:

> Commercial Analytics communicates clearly to businesspeople. They were able to share their methodology with the marketing leaders whose products we are going to be selling. Since those managers are charged with sales force expense, they need to understand and evaluate our pilot.

The strategy execution executive described the communications ability of Commercial Analytics staff in similar terms:

> The members of Commercial Analytics didn't come up through the sales area like I did, but they know they have to translate their findings into something that is "field-friendly." I know the folks in Commercial Analytics are always thinking about how to do that translation. I have worked with analytical people who are much more academic—it is more effective to work with Commercial Analytics.

Embedding Analyses into Tools

One other approach to improving decisions that the Commercial Analytics and Decision Sciences organization takes is to embed results into small software tools for use by marketing and sales managers in the field. The goal is to help field managers make better decisions by providing decision logic and data for the analyses they typically perform.

The group created a "channel choice simulation tool." It allows the user—typically the planner of a marketing campaign—to simulate the decision of channel selection for a particular product. The user can play with a variety of scenarios while attempting to optimize the returns on investments across channels. The output of the simulation is a probability of achieving a certain ROI level for a particular product.

Perhaps the most focused analysis tool is one for sales force vacancy management. If a sales rep leaves a particular region, should the manager fill the vacancy? This tool provides qualitative and quantitative analysis to inform the vacancy-filling decision. In a sense, it's a semiautomated checklist of the factors to consider in filling a sales vacancy. A sales manager's intuitive feeling about the need for a replacement is a key variable in the analysis.

Future Directions for Commercial Analytics and Decision Sciences

The leader of Commercial Analytics and the interviewed clients all feel that the group is providing considerable value for Merck. The key question going forward involves the direction for role expansion. Should Commercial Analytics, for example, expand beyond the U.S. market and provide support for global sales and marketing decisions? Business across Merck has become considerably more global through both acquisitions and organic growth, and the non-U.S. businesses need more analytical help with sales and marketing decisions. The downside, however, would be the possibility of providing too little support for important decisions in the United States, which is the largest market at Merck.

Another option for role expansion would involve more "horizontal" collaboration with other analytics groups across Merck. In addition to Commercial Analytics, Merck has strong analytical capabilities in the R&D/clinical area, as well as in health economics and manufacturing. Thus far, the collaborations among these groups have been relatively minimal. Leadership of Commercial Analytics is aware that some other organizations, both within and outside the pharmaceutical industry, are beginning to view analytics as more of an enterprise-level capability. Thus far, however, the specific benefits of greater collaboration are not clear.

Whatever the future roles of the Commercial Analytics and Decision Sciences organization, the values of independence, clear communications, and assistance to business decision makers in multiple forms will continue. These cultural attributes are an important component of the group's success and have led to a clear competitive advantage for Merck overall and the executives who have taken advantage of the group's capabilities.

Conclusion: Healthcare Analytics: The Way Forward

Dwight McNeill

In this last chapter, I look to the future and discuss some pathways to help the field of healthcare analytics become more effective in improving business and clinical outcomes. I start off by describing the present state of analytics as detailed in the book and conclude that there is much to be done to meet the potential for analytics. I describe the huge agenda, burning platform, and opportunity. I concentrate on the data gold rush and technology as the great enabler and conclude with ways to make change happen.

Analytics As We Know It

This book provides the most comprehensive review of the current state of the art and science of analytics in healthcare including the following:

- A variety of perspectives on the daunting challenges facing the various sectors of the industry, mostly from healthcare reform and hypercompetitive market pressures.

- An overview of healthcare analytic's "DNA": What it is, how it provides value to the business, what are its various forms, what are some examples of its "secret sauce," and how it deals with the ubiquitous pressure of balancing the use of more and more data with the rising concern about personal privacy.

- A workbench of analytics methods including using electronic health records (EHRs) to provide meaningful results; measuring, monitoring, and improving providers' adherence to established clinical standards; reducing medical errors through the use of triggers; finding high cost/clinical need people through "hot spotting"; and using emerging approaches to personalized medicine through integration of genomic and other personal data.

- Examples of best practices through eight case studies on the current state of leading analytics. The common characteristics of these high performing companies are the early adoption and use of EHRs, leadership that clearly articulates organizational mission and goals, the use of clinical warehouses to address organizational needs such as research, the application of analytics to improve business and finance functions, and insights into how to organize analytics for optimal results.

The Gap Between the Cup and the Lip

By all indicators, however, the promise of analytics far overhangs its performance to date to improve clinical and business outcomes. The opportunity number is $300 billion in value annually, according to McKinsey and Company,[1] that is, if "big" data were used creatively and effectively to improve efficiency and quality. That's about one-tenth of the healthcare economy and equal to about one-third of the waste in the health care system. Analytics can actually do better than that. The potential for analytics to make significant contribution to reduce costs, improve the top line, develop new healthcare innovations, and transform the way business is conducted is underrated.

At the present time, the best of what analytics has to offer is practiced by a small minority of healthcare organizations. Indeed it may be as small as a dozen or two. These organizations are large, very well organized—for example, as integrated delivery systems—and have core strengths in traditional forms of analytics, for example, research. This may be a case of the rich getting richer because they can afford the investments in analytics and can see beyond the horizon. The everyday view for most healthcare organizations is a constant pivot to keep up with compliance and regulatory demands, competitor pressures, and legacy systems both technical and cultural.

The majority of healthcare organizations today are still in analytics version 1.0. Their mantra is "don't take my spreadsheets away," and their mindset about analytics is that it is located in the back office somewhere and practiced by statheads who take care of the reporting requirements and do some research types of things and sometimes provide information that is insightful. Many organizations are working hard to build information infrastructure including the EHR, integrate various siloed data warehouses, and upgrade IT capability. This is usually done "over there" in the IT shop, and those in the shop do not want the data "released" until it is housed in the new data warehouse and is clean and perfect. While they are chasing elusive perfection, the clock is ticking on providing value to the business. And there are missed opportunities to demonstrate analytics value with available and relevant data that can be cast into insights for improving the business today. Finally, many entrepreneurial firms are talking about the art of the possible and stressing the opportunities of new analytics innovations to improve the business. Some are building useful solutions.

Overview of the Way Forward

Let's face it. The supply of tools for analytics have been around for quite some time, including deep knowledge on how to collect, integrate, and report data; a large catalog of statistical methods; and a whole field of theory and practice on how to improve decision making. What has changed and *may* offer a tipping point for analytics in healthcare to take off and provide value are three drivers that change the demand for it:

- The pressures from governments and market reforms to change the financing, payment, and delivery of care, along with the clinical information infrastructure (EHR), such that care is less costly, of better quality, more transparent, and responsive to the needs of the retail customers (real people) it serves

- The explosion of data, a data gold rush, in terms of volume and diversity
- Technology advances that significantly improve the cost, time, and capacity to mine, process, and leverage huge volumes of data as well as inverting how data is collected, shared, delivered, and controlled through mobile and biometric devices and social media

But, there is a nagging issue. Making change happen is really very difficult. It's not the tools, data, and technology that matter the most. It's the sociology. John Eisenberg, the esteemed director of the Agency for Healthcare Research and Quality (AHRQ) during its early days, made the point often that if we only translated and applied the research we already knew into practice, it would provide more benefit in terms of improvement in patients' lives than all the research being done to find new causes and treatments.

Indeed the unrelenting focus on technology may be part of the problem. In this era of "bowling alone," the cryptic and hectic release of ideas on Twitter and Facebook, and the wish for frictionless solutions that often preclude engaging with people, we lose sight of the fact that change happens largely through communications among people. According to Atul Gawande, the wish for "turnkey" technical solutions has led us to a preference for "instructional videos to teachers, drones to troops, incentives to institutions."[2] Understanding the technology of diffusing innovations like analytics is just as important as understanding the skills needed for advances in computing such as NoSQL, ETL, and ODS.

So, those in the profession of analytics need to know the business, how analytics dovetails with it, the data, the various computing and analytics technologies, and how to make change happen. That's a tall order. But, the old role does not fit anymore. Let's face it. Most of us practicing analytics are nerds-geeks-statheads who love numbers more than we like people. We like the quietude of the back office much more than the (social) intensity of the front lines. And we are modest and don't fully understand nor can we express our power to make a very big difference in healthcare. To use a phrase from Pogo,

"We have met the enemy and he is us." We need to change the ways we do our job.

A Huge Agenda and Opportunity

U.S. healthcare faces significant challenges and opportunities to improve business and clinical outcomes. Closing the gap on waste and inefficiency can save close to a trillion dollars. If clinical outcomes were improved to match those of the highest performing countries, almost a trillion dollars in value could be achieved in terms of productive lives.[3] And if customer engagement were to match that of the highest performing industries, increased market share could amount to nearly $100 million.[4]

The time is hot for analytics to demonstrate its value. The environmental drivers from government and the marketplace that demand a retail transformation, delivery system changes, payment changes, and a focus on outcomes are all compelling and toughen a burning platform for organizational change...that relies on information...and is fueled by high octane analytics. In addition to these drivers there are other emerging factors to shape the value proposition of analytics:

- **Behavior change**—The pathway to demonstrate value depends on changing behavior across stakeholders, for example, doctors to perform to accepted guidelines for care, business leaders to consider and adopt new innovations, and patients to embrace healthy behaviors. Addressing behavior change at the patient/person/member level is probably most important. This is based on the overwhelming evidence that people's behavior is responsible for good health more than any other determinant and four times more important than healthcare.

- **Data democratization**—Healthcare has been stubbornly provider-centric. But Web 2.0 and social media, mobile devices, and a powerful patient engagement movement are changing how care can be provided, purchased, evaluated, and modified.

- **De-medicalization of healthcare**—It is inevitable that cost pressures and the need for better outcomes will lead to new organizational forms for producing health. For example, medical homes are a great advance, but healthy houses that concentrate on chronic diseases and employ low cost and friendly care by people "just like me" in my home and community are overdue.[5]

- **Predictions**—The availability of huge volumes and different, but highly relevant data, from within and outside healthcare, will make for better decisions especially about the future and grounded in predictive modeling.

We know the usual suspects in terms of where to concentrate and how analytics can contribute. Where to concentrate includes connecting the digital pipes, reducing the voltage drop in the translation of clinical guidelines into practice, measuring and improving the quality and safety of care, supporting healthcare reform, managing population health, and cracking the code on personalized medicine. How to do it with a panoply of analytics methods, theory, frameworks, and best practices has been described in this book. But there is something missing, and it has to do with spread and the bull's eye.

Spread: Spread refers to the diffusion of a practice within and across organizations. For example, when a treatment approach has been proven to have a significant impact on patient outcomes through evidence-based research, one would assume that it would spread like wildfire through healthcare delivery systems. Or when an analytic approach has been teamed with a strategic goal and an implementation plan has been completed down to the last detail, one would expect a quick adoption and flawless implementation. Wrong on both counts! One of the most disappointing features of healthcare for those in the business of improving practice is the s-l-o-w adoption of innovations and low spread. The fact is that change is more complicated than good technology, strategy, incentives, and training manuals. This is discussed in a later section on making change happen.

The Bull's-Eye: Most healthcare change strategies and analytics center on the 2-Ps: providers and payers. But there is a missing P—people. Most of the focus is on health*care*, but healthcare is a small contributor to achieving good health. And, the big, untapped resource for improving health is people. Only they can decide to see a doctor, get insurance, take their medication, follow orders, prevent illnesses through healthy behaviors, decide for themselves what treatments meet their preferences, sort out the providers that are best for them, and use their collective wisdom and power to change the system. The evidence is clear on this point. People control 40% of their health outcomes through their own behaviors.

Another part of the third P is regarding people as customers; that is, knowing them, understanding their needs, anticipating their needs, and serving them in the best definition of customer service.

The last part is understanding that people need to be supported to achieve good health. And it is much more than what happens in the doctor's office. The United States spends more per capita on healthcare, but it spends less than its peer wealthy countries on *health*.[6] Social services matter and, again, are more influential in determining health than is healthcare.

How can analytics serve these ends? Some examples:

- Attack the most dangerous epidemic facing the country— diabetes—by getting outside the box to use extra-industry data to find people with premorbid obesity.

- Engage people in measurement by addressing what matters. What matters is well-being, not whether the hospital is providing beta blockers to heart attack patients in a consistent way. Well-being should be measured, and every municipality and politician should be held accountable for the health of their communities just like they are for education, roads, and jobs.

- Give people dashboards that matter. These are not the self-service kind that instruct people how to fill out their insurance forms. People need compelling tools and active support. They need to be coached on their health journey. Other industries do this well, especially political campaigns.

- They need to be "known" such that services can be radically personalized to their needs. Analytics can provide the tools commonly used in customer analytics to know the person through microsegmentation, tailored messages that are delivered through preferred channels, and interventions that work. This is done in marketing to sell people things. It should be used in healthcare to connect with people to drive behavior change.

- Social supports are critical to health success. Knowing whether a person has financial, nutritional, transportation, and mobility needs can be quantified and are just as important as lab findings.

The bottom line on the way forward is to stay the course on the focus on providers and payers but to widen the aperture to include engagement and interventions with people. It is also important to ratchet up the spread of analytics that produce positive outcomes.

The Data Gold Rush

The data gold rush in healthcare is on. It's the wild, wild West to find and harvest it. We are told that the data produced in U.S. healthcare will soon be counted in yottabytes,[7] or a million trillion megabytes, or 1,000,000,000,000,000,000,000,000 bytes. We are told that creative and thoughtful extraction of all the healthcare big data is worth at least $300 billion a year. Where are all these data coming from, what value is being extracted from them, and where are the untapped opportunities?

Where Is It All Coming From?

The data come from a variety of sources:

- **Transactions**—The traditional sources of usable (structured) healthcare data come mostly from billing (claims) data.
- **Electronic medical records**—EMRs produce useful clinical data in mostly unstructured and semistructured data.
- **Machine-emitted**—Most of the yottabytes come from this source, which includes readings from medical sensors and "scrapings" from Web and social media sources including click-stream and social interaction data.

- **Biometric devices**—These are all the findings from medical measurements such as blood pressure readings and x-rays and other monitors of everything from steps taken (e.g., fitbit) to places visited (GPS).

- **Research**—Data on individuals from clinical trials, registries, and other sources.

- **DNA sequencing**—Genomic data to support personalized medicine are not widely available now but are on the verge of becoming accessible and reasonably priced.

What Value Is Being Extracted From Data?

It is very difficult to know what value is being extracted from these data sources for the purpose of healthcare analytics or for that matter how much money is being spent on healthcare analytics. Most of these data sources were developed for purposes other than healthcare analytics (as defined as providing insights for the enterprise to improve business and clinical outcomes). For example, claims data were developed for billing, and the use of them for understanding clinical analytics has been a stretch. EMRs are for the purpose of improving care and increasing communications among providers. They are not developed and implemented with the notion of combining the data with other sources to get a 360 degree view of the patient or to do comparative effectiveness studies. So, in a way, these data investments have already been made for specific other purposes, as have social media communications and purchases on the Internet. The use of these data for analytics to improve the business or to improve outcomes or develop new products is a secondary use. And the real value add of analytics may be to recombine and repurpose the data. (Later on we discuss the use of data collected primarily for healthcare analytics.)

Let's focus on "big" data and "small" data and hypothesis-driven and hypothesis-free approaches. To start that discussion, let's take a look at mining in another industry: gold.

There are 2,500 metric tons of gold produced annually.[8] At the current price of $1,300 per ounce this amounts to a $100 billion industry. It takes, on average, 30 tons of rock to produce one ounce of gold.[9] Hence, the final product amounts to .000001042 of the rock that needs to be worked through to harvest it. There are also by-products of this extensive mining including the use of cyanide to extract it and huge open pits and large mounds of waste rock across the countryside where it is produced. The mining processes include huge investments in monster shovels and trucks to extract and transport rock to the plant and warehouse for processing and storing. Gold mining is hypothesis-driven; that is, mine rock in a specific place and in a specific way and you get gold. This is quite different from a hypothesis-free approach, which is to take all the rock and do a lot of tests on it to see whether there is anything in it of value.

Yotta-driven analytics in healthcare is mostly hypothesis-free, akin to analyzing the whole mountain and looking to discover "similarities" that may provide new understandings about the delivery of healthcare. The monster computing technology available today can enable seemingly limitless simulations to do this.

If we assume that analytics in all of its permutations in healthcare amounts to just 5% of healthcare spending, this computes to about $135 billion, which is pretty close to the gold mining industry. How much rock will it take to find the gold in healthcare? Will the conversion rate be +/- .000001042?

Some of the gold from the healthcare data rush is palpable and it is "small." The integration of genomic data with clinical data could lead to answers to important questions, such as whether a certain chromosomal variation is related to a disease, which could then fuel individually tailored treatments. For example, Tamoxifen has been an effective drug for the treatment of breast cancer. On average, about 80% of patients benefit from it. The potential with personalized treatment is to become 100% effective in 80% of patients because genetic markers can improve the knowledge of who does and does not benefit from the treatment. There are many instances of "small" hypothesis-driven data that can have a precise impact on business and health outcomes. Other rock in the yotta may not be as clearly useful. For example, much of the yotta is comprised of data emitted from machines, and much more research needs to be conducted to home in on likely ways it can contribute.

Untapped Opportunities

There are two types of data missing from the previous list. These data do not necessarily add a lot to the yotta stats. They are "small" and have specific and targeted purposes. These include extra industry personal data and people-generated data.

Extra Industry Personal Data

The world is full of relevant data and a lot of it resides outside of healthcare. External data can address specific healthcare issues, for example, to change people's behavior, ranging from marketing to early detection of diseases. These data come from privately aggregated and publicly available databases on a wide range of personal attributes that can define microsegments that can be precisely targeted with specific interventions to improve health. For example, data on height and weight are available from external sources (and not easily collected or extracted from usual healthcare data) and can be used to calculate the body mass index (BMI) to determine premorbid obesity. Additionally, when personal data are integrated with medical data and in combination with the right channel—especially mobile—it can produce a much better identification of high-risk patients, with more effective interventions mapped to their specific needs, and include closer monitoring over time.

People Generated Data

Another source of untapped data is people. This is another type of "small data" with big potential benefits. Most of the data sources listed previously do not involve the active participation of people. The real potential lies in gathering much more relevant data from individuals with their consent and engendering their partnership to engage in data-sharing activities that help them improve their life. After all, people know more about their own health and illnesses and can monitor it better than any doctor could possibly hope to do. There is much more to be learned from a person's head than from their data streams. There are indications that this is happening without, and perhaps in spite of, the active strategies of traditional healthcare.

For example, networks of patients with the same condition are sharing data and creating large databases that are beginning to approximate crowd-sourced clinical outcomes research. For example, as of the end of 2011, PatientsLikeMe had more than 120,000 patients in 500 different condition groups; ACOR (Association of Cancer Online Resources) had more than 100,000 patients in 127 cancer support groups; 23andMe has more than 100,000 members in their genomic database. People also engage in their own data sharing through mobile and social media. And people have been responsive to surveys when the purpose is big (like polling in a presidential campaign) and when the rewards for participation are adequate.

Conclusion

A mountain of data is available for analytics in healthcare. Much of it was collected for another function and may be repurposed. Some of it is really big and has unknown uses but is intriguing, and the technology may be able to find the gold although the conversion rate may be infinitesimally small. Some of it is small and can have immediate applications to produce value. And some that is potentially very valuable and comes directly form people is not included in the count and is not collected. Certainly, healthcare lags other industries in its use of big data because of the challenges with complex and unstructured data, the reluctance to use external data, data integration issues, and concerns about patient confidentiality. And IT folks say there is enough unused healthcare industry data to keep them busy for a very long time. Threading the needle for the most productive use of data, whether big or small, hypothesis driven or free, depends on analytics making the case that it is worth the investment and an innovation worth adopting.

Technology, The Great Enabler

Technology is a great enabler for analytics to contribute to the success of an enterprise in the sense that it supplies the computing means and capabilities (hardware, software, and know-how) to solve problems related to data. Moore's law predicted that chip performance would double every two years, which would increase processing speed, memory capacity, sensors, and even the pixels in digital cameras proportionately. For example, comparing the IBM PC released in August 1981 with the Apple iPhone 4 released in June 2010, the CPU clock speed of the PC was 4.77MHz compared to iPhone at 1GHz; the processor instruction size was 16 bits for the PC and 128 bits for the iPhone; the storage capacity of the PC was 160KB and that of the iPhone (base model) was 16GB; and the installed memory (RAM) was 64KB for the PC and 512MB for the iPhone.[10] Additionally, the list price on release of the PC was $3,000 (or about $7,500 adjusted for inflation) and the iPhone was $199, or about 2.5% of the cost of the original PC. This exponential growth in computing performance has driven the impact of digital devices from computers to household appliances in every segment of the world economy.

Has this exponential increase in computing performance had a concomitant increase in information and analytics to improve healthcare? Far from it. One could say that the technology has been ahead of the capacity of organizations to absorb it. It's not the technology that accounts for a slow take-up of analytics, it's the sociology. The hope for technology adoption reminds me of Terrence Mann's prediction about building a ball field in the cornfields of Iowa in the movie *Field of Dreams*: "If you build it, he will come. Ray, people will come, Ray. They'll come to Iowa for reasons they can't even fathom. They'll turn up your driveway not knowing for sure why they're doing it."

In healthcare, organizations do not show up in the driveway to buy technology and do not make large investment decisions without knowing exactly what the ROI is. Technology adoption for analytics in healthcare has not been as rapid as that in science and other industries for a variety of reasons. Healthcare has a cultural underpinning of "do no harm," and it shows in its hesitant approach to change. Why healthcare technology, and analytics innovations in particular, suffer a slow pace of adoption is addressed in the next section on making change happen.

The technology breakthroughs available to healthcare are awesome and can make analytics quicker, cheaper, and smarter. A few are discussed briefly in the following sections.

NoSQL (Not Only SQL)

These databases are an alternative to traditional, relational databases and are especially suited for unstructured big data, Web 2.0, and mobile applications. It uses open source software that supports distributed processing. It scales "out" to the cloud, rather than "up" with more servers. It has fewer data model restrictions than relational databases management systems, which allows more agile changes and less need for database administrators. It can use low cost commodity hardware. The bottom line is that it is faster and much cheaper. Examples of popular NoSQL databases include Cassandra, Hadoop, and BigTable. Companies that use it include Facebook, Netflix, LinkedIn, and Twitter. For more information see the NoSQL website, which touts itself as "your ultimate guide to the non-relational universe."[11]

High Performance Computing (HPC) Through the Cloud

High performance computing (HPC) allows users to solve complex science, engineering, and business problems using applications that require high bandwidth, low latency networking, and very high compute capabilities. This is the computing capability needed for mining mountains of data. This capacity can be provided by dedicated computer clusters or by cloud clusters. Dedicated, custom-built, supercomputer infrastructure requires significant capital investments, long procurement times, long queues, and extensive database management. Buying HPC services from the cloud provides definite cost advantages, short lead teams, access to the scale required for a given project, and on-demand capacity. An example of such an offering is from Amazon Web Services called Cluster Compute Instances.[12] In healthcare, the biopharma sector uses HPC for genome analysis. Other industries, including oil and gas, financial services, and manufacturing, use it for modeling.

Machine Learning

The idea that machines could replace humans for certain functions has been around a long time. And it certainly has become commonplace in industries such as automotive with robots on the assembly line. But can the machines actually "learn" and improve functioning on their own beyond being explicitly programmed? There are good examples of this with Google Search and Amazon purchasing recommendations, and with voice and facial recognition applications. In healthcare, IBM demonstrated a compelling use of machine learning (and natural language processing and predictive analytics) with its Watson technology by beating two grand champions on the *Jeopardy!* TV quiz show. IBM is now moving beyond quiz shows and working on healthcare solutions, mostly in the area of differential diagnosis. One of the institutions it has partnered with is Memorial Sloan-Kettering Cancer Center.[13] The goal according to Sloan-Kettering is to have the technology gather and assimilate information from the research literature and from the Center's clinical experience documented in its medical records and other files to "bring up-to-date knowledge to the bedside of every cancer patient."[14] Watson might be able to do this through its capabilities to read and understand language, interact with humans, remember everything, and provide answers to real-time questions. How the information will be delivered to the physician, how it might transform the practice of medicine, and whether physicians will embrace the technology are all important, open questions. It is certainly a very bright star illuminating and guiding the emerging field of machine-enabled clinical decision support.

Clicks and Mobile

The Internet has transformed the way businesses communicate, market, do commerce with customers, and collect data about them. In retail, clicks are challenging the bricks. What could be more indicative of shifting paradigms than the collapse of the structures in which people do business (stores). One example is the capability to do randomized trials, or A/B testing, of alternative Web site features—for example, how to get the most contributions during a political campaign—on large samples and virtually instantaneously. Another example is Web page "scraping" in which all types of data about people's Web wanderings are turned into ratings about their suitability for a job, a loan, and a date.

More than half of the adult population in the United States have smartphones.[15] Facebook has more than 1 billion monthly users.[16]

The hot combination of these two popular technologies produces a platform for easy, convenient, and quick communications that also enable e-commerce, uber-targeted marketing, location monitoring, and much more. An opportunity going forward in healthcare is to create closer relationships with people to help them get healthier by tapping into data that are freely exchanged and by supporting the continual, fast evolution of new applications to support health.

The Two Faces of Enabling

Technology can enable analytics in two ways. The positive face of enabling is seen in its tremendous success story in increasing computing capacity with hardware (speed, memory, storage, access, etc.) and with software (to manage all the data and make sense of it). But technology is mostly content-agnostic. It really does not care about the content or use of the data. It is a template for manipulating it. And often, it perpetuates GIGO, or garbage in/garbage out. That is, if the data, models, and assumptions that are foisted on the computer by humans are wrong, the computer cannot correct them. And now, with the capacity to process more data more quickly, there is the possibility of more garbage.

The flip side of enabling is to keep people from experiencing and learning about challenges and their consequences by unwittingly helping them, but in the wrong way. This may serve to protect the individual from harm but actually exacerbates the discovery of new solutions. Technology can be so seductive to analysts that it blinds them from keeping their eyes on the prize. The prize is changing behavior to improve health and reduce costs. And with most of these challenges in healthcare, there is an adequate supply of technology. What's missing is the time and attention it takes to make change happen with people, not machines.

Making Change Happen

With the scorching, burning platform for healthcare transformation; the availability of huge volumes and diverse types of data; technologies to make computing quicker, better, and cheaper; and hype about the art of the possible with big data; one might ask why the uptake of healthcare analytics is so low in comparison to its potential and relative to the performance of other industries? In addition to the usual attributions including the need to perfectly digitize the industry, there are industry context barriers that deter innovation, including lack of good coordination, physician autonomy, convoluted market dynamics, multiple vested and powerful interests, and a pervasive, risk averse culture. But, I would like to concentrate on two possible answers and solutions that analysts can and should address: The technology of change is underrated, and analysts must change their understanding of their role and contribution.

Technology of Change

Making change happen is not as rational as we analysts would hope. We assume that if we demonstrate the ROI for an innovation and put it in a good business plan that decision makers will burn a path to our door to implement our recommendations. It's hard enough to do this for conventional business functions but harder for analytics innovations. Making change happen is much more complicated and subtle, and much less about numbers than about people.

Healthcare works really hard to change people's behavior. It borrows from aviation to have pre- (flight) surgical checklists to reduce mistakes like wrong leg amputations, it issues alerts (continuously), for example, when the wrong medication is ordered on the CPOE (Computerized Physician Order Entry) system, it measures and publicly reports on the performance it wants because it believes what is measured is managed well, and it conducts a lot of research to get the evidence on what works and what doesn't to improve the knowledge base and shape the delivery of care.

Most strategies to get people to change the way they do things often boil down to three approaches: Please do it and I will teach you how; you must do it and I will punish you if you don't; and do it well and I will pay you a (modest) reward for doing so. These are all rational approaches and make sense and are a part of the armamentarium of change....And they are insufficient.

What we are trying to do with analytics is embed them in the daily operations of an organization such that it becomes the normative way of doing things. More than that we want the analytics to be successful and produce the intended results on an assured, repeatable basis. Eventually we want the analytics innovation to give way to its own reinvention as it adapts to the needs of an organization. This is a long journey and cannot be short circuited. It involves mastering six stages:

1. **Ideas**—Ideas are the starting point and backbone of innovations. President John F. Kennedy exalted ideas. He said, "A man may die, nations may rise and fall, but an idea lives on."

2. **Design**—Ideas need to be converted into a theory of action on how it will accomplish a goal.

3. **Decision**—This is when the design is proposed to decision makers, and it receives an up, down, or delay action. Most ideas do not pass the test for adoption. Machiavelli said, "There is nothing more difficult to plan, more doubtful of success, nor more dangerous to manage than the creation of a new order of things." More on this later.

4. **Implementation**—Implementation is about following the design rule book in carrying out all the required process steps. Goethe noted in the eighteenth century, "To put your ideas into action is the most difficult thing in the world."

5. **Evaluation**—Stakeholders need to understand performance to make adjustments in the design and operations so that it becomes institutionalized, altered to fit the changing needs of the organization, or discontinued. More often than not, ideas get stuck in the pathway and do not live to fight the next stage. And implementations of complex programs suffer from innumerable snags in delivery and most fail.

6. **Reinvention**—Reinvention is important and is not a failure of intent to implement a plan "as written." For example, one might think that clinicians would adopt clinical guidelines developed by their professional organizations as is. But success might depend on not getting the full loaf and instead adopting a local version that works in a particular context. Hopefully, all innovations evolve and change through a learning and improvement process.

Analytics cannot be responsible for all elements of the innovation pathway. After all that is what operations folks and c-suite leadership are about. But what analysts must do is master the first three stages including ideas, design, and decision. Analysts must know how to get their projects known and funded.

The technology of adoption is complex and multifaceted. Adoption is about making a decision to activate an innovation into practice. The process of adoption is largely about collecting, processing, and evaluating information to understand the innovation and to reduce uncertainty in relation to its pros and cons. A useful model based on the works of multiple scholars of making change happen is displayed in Figure C.1. The model includes six domains consisting of 18 factors which need to be addressed for the successful adoption of an innovation:

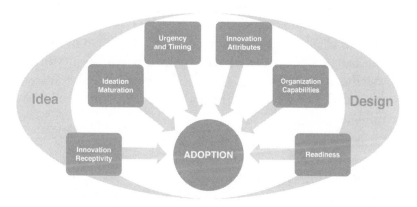

Figure C.1 Innovation adoption factors model

(Source: Dwight McNeill, *A Framework for Applying Analytics in Healthcare: What Can Be Learned from the Best Practices in Retail, Banking, Politics, and Sports.*[17])

- **Innovation receptivity**—This includes the important consideration of receptive context, that is, whether the land (organization) is fertile, arable, and moist enough to grow seed. To what extent is the organization receptive to new ideas? Does it nurture new ideas or penalize the troublemakers who question the status quo? Are there survival pressures that require change such as the drivers stated earlier? What are the existing norms and beliefs about change in general? Is the culture rooted in values of "better to be safe than sorry" and "do no harm" and conservative in taking on risks?

- **Idea maturation**—This includes whether the idea is articulated sufficiently and is ready for prime time. Has it confronted confirmation bias and looked beyond "seeing only the things we are looking for"? Is it clear about the cause and effect theory? If this is not clear, and what the steps are, the argument can shift to ideological and political issues.

- **Urgency and timing**—Is there a burning platform for change that the idea is associated with and is there a window of opportunity within which all the stars must get aligned?

- **Innovation attributes**—The innovation attributes domain is important and includes the key factors of the relative advantage of the innovation, its compatibility with existing processes and attitudes, its complexity, and its "trialability." Of most importance is whether the innovation is judged to be better than an alternative. What is its ROI? Is it compatible and fits in with the way the organization does things? How complex is it to understand and use? Can it be given a test drive before a full scale roll-out?

- **Organization capabilities**—Who is making the decision to adopt—an individual, team, or ultimately the users who can support or torpedo it? Is the program component of the analytics innovation capable of making it work? For example, analytics can precisely segment people into their risk for morbid obesity. But is there a compelling theory about how program staff actually change the behaviors of those identified? Similarly, does the organization have the people skills and technologies to implement the innovation successfully?

- **Readiness**—This includes evidence from small scale testing of an innovation to show its impact and communications to position the innovation favorably such that those who are expected to deploy it are ready and willing. And last but not least is leadership ready and willing to support it? On the last point: Leadership at the executive level is about creating the vision and executing a strategy to "get people from where they are to where they have not been," in the words of Henry Kissinger. Succeeding at innovation requires leadership that creates an environment that welcomes and encourages new ideas and change, cuts through the confusion and uncertainty to be decisive, and then leads the organization to execute flawlessly. Ideally, the organization is a learning system that is constantly updating its mental models and making improvements. Without this executive leadership, the likelihood of adoption of new innovations is diminished.

In summary, there are six stages of the innovation pathway to success. For one of these stages, decision, there are six domains and 18 factors. For the other stages, there is the same degree of complexity along with clear frameworks for addressing it. So, there is a technology of change that is well articulated.

Atul Gawande, in his essay in the *New Yorker*, "Slow Ideas,"[18] boils this all down to a simple suggestion. He reiterates the wisdom of Everett Rogers[19] from 50 years ago that "Diffusion is essentially a social process through which people talking to people spread an innovation." Gawande presents a case study on how to get hospitals and birth attendants in India to carry out a few of the tasks required for safer childbirth, such as washing hands, keeping the baby warm, and taking vital signs properly. He asked a nurse why she changed her behavior and how it became something she did "day in and day out, even when no one is watching." She said that she did not listen to the teacher at first because of her heavy workload, but that after multiple visits with the teacher/coach she began to listen and change. Asked why, she said the teacher was "nice." "It wasn't like talking to someone who was trying to find mistakes. ...It was like talking to a friend." He poses the question, "So, what about just working with healthcare workers, one by one to do just that?" Sounds corny? A fundamental, yet overlooked, cornerstone of making change happen is people-to-people communications, done often, and with great respect toward the individual's existing norms and beliefs.

"Be the Change..."

If change is not optimal in healthcare and analytics is not precipitating it, then something has got to give. In this section I address the field of analytics professional identity, salvation by the Chief Analytics Officer (CAO), and our individual role to "be the change."

Identity Crisis

Who are we (as analysts) and what do we do? What is our identity? Identity is defined as the state or fact of remaining the same one under varying aspects or conditions. How do analysts define themselves? The words that come to mind most often include quant, stathead, and data scientist. How we define ourselves reflects our purpose and what work we do.

What do we do and how do we contribute (functions)?

- Create infrastructure for generating and analyzing data.
- Produce platforms for data sharing.
- Create one version of the truth.
- Adopt the best information technologies.
- Keep information infrastructure costs low.
- Produce standard reports and ad hoc queries.
- Produce products that can support decision making.
- Execute information management strategies.
- Work collaboratively with leaders across the enterprise to support their achievement of strategic priorities.
- Provide insights that are valuable to the business.
- Contribute to the growth of the business by demonstrating a contribution to the earnings per share.

All of these are relevant to what analysts do. Those at the top of the list are more common today. Those at the bottom are emerging as the field advances in its development and focuses on the ends (outcomes) more than the means.

What do we need to know to contribute to purpose (competencies)?

- Data
- Computing and analytics methods and technologies
- The nature of the business and knowledge about its customers
- How analytics creates value
- How to negotiate change in an organization
- How to be a leader

The first two competencies are what we most often think of when we think of analytics. Our contributions get more significant as we go down the list and ratchet up our engagement in making the business work better and grow.

So, our identity is changing, hopefully, as we advance in the maturity of our profession and change the scope of our work from doing things with data to using data to change the organization for the better.

But this is an overwhelming list of functions and competencies. No one person can do it all. It requires a village (team) to have all the skills and accomplish all the functions. And some suggest that a new chief is needed.

Salvation from the Chief Analytics Officer (CAO)?

CAO stands for chief analytics officer. (It also stands for chief administrative officer.) The role is new for the C-suite. Michael Bloomberg, the data-driven mayor of New York City, in his last State of the City Address,[20] appointed the city's first ever CAO, Michael Flowers, to improve the way all agencies share information and to make the data available to the public so that the community can hold the city accountable. Flowers used to be the city's analytics director. It's not clear why there was a title change. The job description sounds better in the previous job. "Mr. Flowers leads a team of data scientists in analyzing city data from over 20 city agencies to allocate its resources quickly and efficiently to prevent fire, crime, safety hazards, and unhealthy conditions."[21]

Chiefs are becoming very popular. Forbes lists new C-suite titles[22] including chief Internet evangelist, chief happiness officer, chief privacy officer, chief digital officer, chief knowledge officer, and chief customer officer among others. In healthcare, the new CAO role joins forces with other information leaders including the CIO (information), CDO (data), CIO (innovation), CMIO (medical informatics), and CNIO (nursing informatics). I bet there are more to come.

I guess the reason for chiefs is to bring visibility to the function, get the ear of the CEO, collaborate with peer chiefs for the good of the enterprise, provide better management oversight, and be accountable for results. All good things, and it's important that analytics is recognized as an important function along with the dozens of others. It is good to have the executive talent, but leadership is not just for the few chiefs. It's for all of us.

Remake Ourselves

Mahatma Ghandi said "Be the change that you wish to see in the world." We need to expand our technical and people skills to increase the utility of analytics in healthcare. We need to work locally and make teams work better through communications and collaboration and dedication to a common goal. We need to focus on the immediate tasks at hand, such as working through an algorithm or building a database and also be sure there is a receptor site to absorb our work. We need to visualize how analytics improves business and society. Ultimately, we need to lead by our own example.

Notes

1. James Manyika et al. Big Data: The Next Frontier for Innovation, Competition, and Productivity. McKinsey Global Institute, May 2011. http://www.mckinsey.com/insights/mgi/research/technology_and_innovation/big_data_the_next_frontier_for_innovation

2. Atul Gawande, "Slow Ideas," *New Yorker* (July 29, 2013), p. 45.

3. Dwight McNeill, *A Framework for Applying Analytics in Healthcare: What Can Be Learned from the Best Practices in Retail,*

Banking, Politics, and Sports (Upper Saddle River, NJ: FT Press, 2013).

4. Ibid. p.31.

5. Suzy Hansen, "What Can Mississippi Learn from Iran," *New York Times*, July 27, 2012.

6. Elizabeth Bradley and Lauren Taylor, "To Fix Health, Help the Poor," New York Times, www.nytimes.com/2011/12/09/opinion/to-fix-health-care-help-the-poor.html??_r=0

7. P. Cerrato, "Is Population Health Management the Latest Health IT Fad?" *Information Week* (July 31, 2012) retrieved December 12, 2012, http://www.informationweek.

8. "All the World's Gold," http://www.numbersleuth.org/worlds-gold/.

9. "Behind Gold's Glitter: Torn Lands and Pointed Questions," *New York Times*, June 14, 2010, http://www.nytimes.com/2005/10/24/international/24GOLD.html?pagewanted=all.

10. Alfred Poor, "30 Years Later: An Unfair Comparison Between an IBM PC and an Apple iPhone 4," HP.com, December 4, 2012, http://h30565.www3.hp.com/t5/Feature-Articles/30-Years-Later-an-Unfair-Comparison-between-an-IBM-PC-and-an/ba-p/2662.

11. NoSQL.org, http://nosql-database.org/.

12. Amazon Web Services, What's New?, "Announcing High Memory Cluster Instances for Amazon EC2," http://aws.amazon.com/about-aws/whats-new/2013/01/21/announcing-high-mem-cluster-instances-for-amazon-ec2/.

13. IBM, "Memorial Sloan-Kettering Cancer Center, IBM to Collaborate in Applying Watson Technology to Help Oncologists," March 22, 2012, www-03.ibm.com/press/us/en/pressrelease/37235. wss.

14. Larry Norton, "How Collaboration Between IBM and Memorial Sloan-Kettering Taps the Wisdom of Physicians," *Huffington Post*, Blog, March 29, 2012. www.huffingtonpost.com/dr-larry-norton.

15. Mark Rogowsky, "More Than Half of Us Have Smartphones Giving Apple and Google Much to Smile About," Forbes.com, June 6, 2013, http://www.forbes.com/sites/markrogowsky/2013/06/06/more-than-half-of-us-have-smartphones-giving-apple-and-google-much-to-smile-about/.

16. Donna Tam, "Facebook by the Numbers:1.06 Billion Monthly Active Users," CNet.com, http://news.cnet.com/8301-1023_3-57566550-93/facebook-by-the-numbers-1.06-billion-monthly-active-users/.

17. Dwight McNeill, *A Framework for Applying Analytics in Healthcare: What Can Be Learned from the Best Practices in Retail, Banking, Politics, and Sports* (Upper Saddle River, NJ: FT Press, 2013).

18. Atul Gawande, "Slow Ideas," *New Yorker*, July 29, 2013.

19. Everett Rogers, *Diffusion of Innovations*, first edition (New York: Free Press, 1961).

20. "Mayor Bloomberg Delivers 2013 State of the City Address," NYC. gov, Press Release, February 14, 2013, online at http://www.nyc.gov/portal/site/nycgov/menuitem.c0935b9a57bb4ef3daf2f1c701c789a0/index.jsp?pageID=mayor_press_release&catID=1194&doc_name=http%3A%2F%2Fwww.nyc.gov%2Fhtml%2Fom%2Fhtml%2F2013a%2Fpr063-13.html&cc=unused1978&rc=1194&ndi=1.

21. "Michael Flowers," Champions of Change, The White House, http://www.whitehouse.gov/champions/local-innovation/michael-flowers.

22. "The New C-Suite Titles," Forbes.com, http://www.forbes.com/pictures/fhgl45eglf/chief-internet-evangelist/.

Index

A

ACA (Affordable Care Act), 78, 121

academic medical centers, analytical capabilities, 17

acceptable use, information security, 99

accountability for measures, 136

Accountable Care Organizations. *See* ACOs

ACOs (Accountable Care Organizations), 4, 21, 121, 145, 270

action provided by accountable stakeholders, 137

actionability, business value of analytics, 84-85

actualizing value management, business value of analytics, 89-90

actuarial rating, 25

adaptations, cross-pollination across industries, 105-114

adoption of best practices, 187-193

  BI performance analytics, 87

  life sciences organizations, 192-193

    Merck, 273-279

payer organizations, 190-192

  Aetna, 245-250

  EMC Corporation, 253-271

provider organizations, 188-190

  AFMS, 231-238

  Catholic Health Initiatives, 213-220

  HealthEast Care System, 239-243

  Partners, 195-211

  VHA, 223-230

advanced analytics, healthcare improvements, 161-167

  care management expert system, 164

  hot spotting, 161-163

  multivariate analytics, 165-167

Aetna, 245-250

  analytical capabilities, 29

  bellwether lessons, 249-250

  history, 246

  Maturity Model, 248-249

  organization, 246-248

affinity group sites, 176

Affordable Care Act. *See* ACA

W